T0292170

Ketamine for Treatment-Resistant Depression

Ketamine for Treatment-Resistant Depression

Neurobiology and Applications

Edited by

Gustavo H. Vazquez

Professor
Lead, Ketamine Clinic, Mood Disorders Outpatient Unit
Queen's University, Department of Psychiatry
Providence Care Hospital
Kingston, ON, Canada

Carlos A. Zarate

Chief Experimental Therapeutics & Pathophysiology Branch &
Section Neurobiology and Treatment of Mood Disorders
Division of Intramural Research Program
National Institute of Mental Health
Bethesda, MD, United States

Elisa M. Brietzke

Professor
Kingston General Hospital
Providence Care Hospital
Queen's University School of Medicine
Kingston, ON, Canada

Academic Press is an imprint of Elsevier
125 London Wall, London EC2Y 5AS, United Kingdom
525 B Street, Suite 1650, San Diego, CA 92101, United States
50 Hampshire Street, 5th Floor, Cambridge, MA 02139, United States
The Boulevard, Langford Lane, Kidlington, Oxford OX5 1GB, United Kingdom

Notices

Knowledge and best practice in this field are constantly changing. As new research and experience broaden our understanding, changes in research methods, professional practices, or medical treatment may become necessary.

Practitioners and researchers must always rely on their own experience and knowledge in evaluating and using any information, methods, compounds, or experiments described herein. In using such information or methods they should be mindful of their own safety and the safety of others, including parties for whom they have a professional responsibility.

To the fullest extent of the law, neither the Publisher nor the authors, contributors, or editors, assume any liability for any injury and/or damage to persons or property as a matter of products liability, negligence or otherwise, or from any use or operation of any methods, products, instructions, or ideas contained in the material herein.

Library of Congress Cataloging-in-Publication Data
A catalog record for this book is available from the Library of Congress

British Library Cataloguing-in-Publication Data
A catalogue record for this book is available from the British Library

ISBN: 978-0-12-821033-8

For information on all Academic Press publications visit our website at https://www.elsevier.com/books-and-journals

Publisher: Nikki Levy
Acquisitions Editor: Melanie Tucker
Editorial Project Manager: Kristi Anderson
Production Project Manager: Sreejith Viswanathan
Cover Designer: Alan Studholme

Typeset by TNQ Technologies

Contents

Contributors ... ix

CHAPTER 1 Ketamine, Clio, and the hippocratic triangle—fragments of the history of ketamine ... 1

Casimiro Cabrera-Abreu, LMS, MSc, MRCPsych, FRCPC and Mariel Cabrera-Mendez, MD

Introduction ... 1
A note on methodology ... 2
The origin of the term dissociative anesthesia ... 3
A note on the history of dissociation ... 5
From Phencyclidine and ketamine to model psychosis ... 6
Dumb me, dumb me! ... 7
Final reflections ... 9
References ... 10

CHAPTER 2 Ketamine's potential mechanism of action for rapid antidepressive effects – a focus on neuroplasticity ... 13

Melody J.Y. Kang, BScH

Background ... 13
Ketamine restores molecular neuroplastic molecules to induce rapid antidepressant effects ... 14
Glutamate, glutamine, and γ-aminobutyric acid availability and cycling ... 15
Enhanced activation of the α-amino-3-hydroxy-5-methyl-4-isoxazolepropionic acid receptor ... 16
Increased mechanistic target of rapamycin signaling pathway activation ... 16
Inhibition of MAPK/ErK ... 20
microRNA expression ... 21
Divergence of mechanisms for ketamine's rapid effects versus sustained effects ... 21
Limitations to the molecular neuroplasticity mechanism and recent findings ... 22
Conclusion ... 24
References ... 24

CHAPTER 3 Treatment resistant depression 33
Sophie R. Vaccarino, HBSc and Sidney H. Kennedy, MD
Defining treatment-resistant depression..33
Staging models...34
Comparing staging models 37
Prevalence..37
Diagnosis and differential diagnosis...38
Etiopathology..39
Treatment strategies 41
Switching antidepressants..42
Combination and adjunctive therapies 45
Neuromodulation 52
Psychotherapy...57
Psychedelics 57
N-methyl-D-aspartate receptor antagonists: ketamine and esketamine 59
Conclusion 60
References..61

CHAPTER 4 Suicide in psychiatric disorders: rates, risk factors, and therapeutics 85
Leonardo Tondo, MD, MSc and Ross J. Baldessarini, MD
Introduction..85
Suicidal risks in the general population 85
Risk factors 86
Suicidal risks in psychiatric disorders.....................................86
Risk factors for suicidal behavior 87
Therapeutics 90
Challenges for studies of suicide prevention............................90
Antidepressants 92
Lithium...94
Anticonvulsants...96
Antipsychotics 96
Anxiolytics and sedatives...98
Ketamine...98
Psychotherapy and other interventions.................................. 100
Conclusions.. 100
Acknowledgments... 104
References.. 104

CHAPTER 5 Overview of ketamine for major depression: efficacy and effectiveness 117

Anees Bahji, MD, Gustavo H. Vazquez, MD, PhD, FRCPC, Elisa M. Brietzke, MD, PhD and Carlos A. Zarate, MD

Historical background... 117
The antidepressant properties of ketamine: evidence from animal studies 118
Open-label and nonrandomized studies on the antidepressant potential of ketamine... 119
Ketamine for suicidality... 119
Evidence from randomized controlled trials 120
Role of ketamine in electroconvulsive therapy for depression....... 120
Possible mechanisms for the rapid antidepressant effects of ketamine 121
Predictors of treatment response to ketamine............................ 122
Limitations and caveats 123
Conflict of interest statement 123
References... 124

CHAPTER 6 How to implement a ketamine clinic................ 131

Ranjith Chandrasena, MD, MSc, Jonathan Fairbairn, BMSc, MD, Melody Kang, MD, MSc and Gustavo H. Vazquez, MD, PhD, FRCPC

Background 131
Administrative approval... 131
Initiation of the clinic... 132
Patient evaluation 133
The clinical setting for ketamine 134
Treatment protocol... 134
Ketamine infusion frequency—acute and maintenance therapies 135
Sustainability of the infusion service....................................... 136
Conclusion 137
References... 137

CHAPTER 7 Development of new rapid-action treatments in mood disorders 139

Elisa M. Brietzke, MD, PhD, Rodrigo B. Mansur, Fabiano A. Gomes, MD, PhD and Roger S. McIntyre, MD, FRCPC

Introduction... 139
Molecular targets for rapid-action treatments............................ 140
  CLOCK genes... 140

Glutamatergic transmission 140
Cholinergic transmission 141
Overview of underdevelopment rapid-action treatments 142
Esketamine 142
Rapastinel 142
Scopolamine 143
Conclusions 143
References 144
Further reading 146

CHAPTER 8 Closing remarks 147

Elisa M. Brietzke, MD, PhD, Carlos A. Zarate, MD and Gustavo H. Vazquez, MD, PhD, FRCPC

Introduction 147
State of the art and current challenges 147
Limitations of knowledge 148
Challenges in knowledge translation 149
Conclusion 149
References 150

Index 153

Contributors

Anees Bahji, MD
Resident, Department of Psychiatry, Queen's University School of Medicine, Kingston, ON, Canada; Doctor, Psychiatry, Department of Public Health Sciences, Queen's University, Kingston, ON, Canada

Ross J. Baldessarini, MD
Research Associate, Psychiatry, McLean Hospital, Belmont, MA, United States; Professor, (Neuroscience), Harvard Medical School, McLean Hospital, Belmont, MA, United States

Elisa M. Brietzke, MD, PhD
Professor, Kingston General Hospital, Providence Care Hospital, Queen's University School of Medicine, Kingston, ON, Canada

Casimiro Cabrera-Abreu, LMS, MSc, MRCPsych, FRCPC
Associate Professor, Psychiatry, Queen's University, Kingston, ON, Canada; Attending Psychiatrist, Mood Disorders Research and Treatment Service, Providence Care Hospital, Kingston, ON, Canada

Mariel Cabrera-Mendez, MD
Research Assistant, Providence Care Hospital, Kingston, ON, Canada

Ranjith Chandrasena, MD, MSc
Professor, Chatham-Kent Health Alliance, Chatham, Ontario, Canada

Jonathan Fairbairn, BMSc, MD
Professor, Chatham-Kent Health Alliance, Chatham, Ontario, Canada

Fabiano A. Gomes, MD, PhD
Assistant Professor, Department of Psychiatry, Queen's University School of Medicine, Kingston, ON, Canada; Kingston General Hospital, Kingston, ON, Canada

Melody J.Y. Kang, BScH
Master of Science, Centre for Neuroscience Studies, Queen's University, Kingston, ON, Canada

Sidney H. Kennedy, MD
Dr. Psychiatry, University Health Network, Toronto, ON, Canada

Rodrigo B. Mansur
Department of Psychiatry, University of Toronto, Toronto, ON, Canada; Mood Disorders Psychopharmacology Unit (MDPU), University Health Network (UHN), Toronto, ON, Canada

Roger S. McIntyre, MD, FRCPC
Department of Psychiatry, University of Toronto, Toronto, ON, Canada; Mood Disorders Psychopharmacology Unit (MDPU), University Health Network (UHN), Toronto, ON, Canada; Brain and Cognition Discovery Foundation (BCDF), Toronto, ON, Canada

Leonardo Tondo, MD, MSc
Director, Psychiatry, Mood Disorder Lucio Bini Center, Cagliari, Italy; Research Associate, Psychiatry, McLean Hospital, Belmont, MA, United States; Professor, (Neuroscience), Harvard Medical School, McLean Hospital, Belmont, MA, United States

Sophie R. Vaccarino, HBSc
Research Assistant, Centre for Depression and Suicide Studies, Unity Health Toronto, Toronto, ON, Canada; MSc Candidate, Institute of Medical Science, University of Toronto, Toronto, ON, Canada

Gustavo H. Vazquez, MD, PhD, FRCPC
Professor, Lead, Ketamine Clinic, Mood Disorders Outpatient Unit, Queen's University, Department of Psychiatry, Providence Care Hospital, Kingston, ON, Canada

Carlos A. Zarate, MD
Chief, Experimental Therapeutics and Pathophysiology Branch, Section Neurobiology and Treatment of Mood Disorders, Division of Intramural Research Program, National Institute of Mental Health, Bethesda, MD, United States

CHAPTER

Ketamine, Clio, and the hippocratic triangle—fragments of the history of ketamine

1

Casimiro Cabrera-Abreu, LMS, MSc, MRCPsych, FRCPC [1,2], Mariel Cabrera-Mendez, MD [3]

[1]*Associate Professor, Psychiatry, Queen's University, Kingston, ON, Canada;* [2]*Attending Psychiatrist, Mood Disorders Research and Treatment Service, Providence Care Hospital, Kingston, ON, Canada;* [3]*Research Assistant, Providence Care Hospital, Kingston, ON, Canada*

Introduction

In 2003, Edward Shorter and Peter Tyrer, a historian of psychiatry and one of the associate editors at the time of the British Journal of Psychiatry, respectively, published an intriguing paper about the nature of "cothymia" and the accompanying drought in the discovery of new antidepressants.[1] Their paper was boldly published during a period when the hegemony of the *Diagnastic and Statistical Manual of Mental Disorders* (*DSM*) classification system was undisputed and absolute; nobody who wished to publish in an August psychiatric journal dared not to use *DSM*. Tyrer, with his characteristic dry wit, touched upon this issue in a later paper with a similar subject.[2] By the beginning of the next decade, Stephen E. Hyman[3] deplored the announcement of a number of pharmaceutical companies concerning the closure of several lines of investigation in the field of psychotropic medications; the stagnation in the production of novel antidepressants became alarming. It is therefore remarkable that in a time when the future of the psychopharmacology of depression appeared to be barren, a "revival" of an "old" substance from the field of anesthesia appeared to "revolutionize" the treatment of depression.[4] Ketamine was introduced in anesthesia about the time the first "antidepressants" were launched (the 1950 and 1960s) but had to wait 50 years to become the new enfant terrible of psychopharmacology.

In a recent paper, John Krystal and his collaborators,[4] incorporating the "revolutionary" halo carried by ketamine, hailed the introduction of ketamine for depression as a proverbial "paradigm shift," *á la Kuhn*,[5] in psychopharmacology, comparable to the "Psychopharmacological Revolution" of the aforementioned 1950 and 1960s. This chapter covering the (admittedly fragmented and partial) history of ketamine unravels thus in the context of the triumphalist assumptions of the turn-of-the-century diagnostic psychiatry[6] culminating in the recent controversies (debacle?)

Ketamine for Treatment-Resistant Depression. https://doi.org/10.1016/B978-0-12-821033-8.00001-0

of *DSM-5*.[7] The issue of the sociohistorical context is apposite and with ironic justice brings to mind the image of the "set and setting" of Hartogsohn.[8]

The purpose of this chapter is to follow the tribulations of ketamine and some of its psychotropic effects and how they have been interpreted in the context of the changing nosologic and nosographical vicissitudes of North American psychiatry during the past 5 decades. Without doubt, Edward Domino has been an important figure throughout this period; some of his encounters with ketamine and its congeners will also be briefly reviewed.

A note on methodology

The issue of context highlighted earlier—where and when does this narrative take place—leads us to the problem of the historiographical (i.e., the methodology) tools used in this chapter. This is important because there appears to be a small cottage industry rapidly building up around the history of ketamine.[9–12] Although essential reading, it is difficult not to think of Ben Shephard's[13] comments when describing the "functional approach to history" of busy psychiatrists [and academic researchers] at the time of writing the history of posttraumatic stress disorder, which according to him is comparable to that of the Communist Party in The Brezhnev epoch, "The medical literature of the past is important and interesting when it buttresses and legitimizes present practice. When it doesn't, forget it".[13] Be as it may, the historical snapshots offered in this chapter supplement the contributions of those authors.

Since Kuhn's weaponization of Butterfield's concept of 'whiggishness' as a fundamental principle in the history of science,[14] a "presentist, internalist, or hagiographic" approach/methodology has been declared deficient or suboptimal[15] at the time of pursuing historical endeavors. This creates significant problems for the two authors, a clinical psychiatrist and an MD, none of whom have professional training in the methods of 'History' (with a capital H), who attempt to operate within the current fashion of the rigorous bounds imposed by the 'evidence-based' study of facts. The temptation of writing a history with a 'king and battle bent,' as Edwin Wallace IV once said[16] or *á la Brezhnev*, in the words of Ben Shephard, is considerable. One of the approaches followed in this chapter is inspired by the 'Hippocratic triangle' posited by Jacalyn Duffin[17] that includes the 'clinician-historian' and its study of history as an inherently flexible task or a set of tasks in which regular calibration and recalibration, from several sources, should be explicitly stated (which is also redolent of the methodology of Berrios). Needless to say, other subtle influences are at work when writing these lines. We can only hope that they will become more or less visible as the text unravels; that is, the intended purpose is that these 'methodologies' and influences will become evident when addressing different aspects of the history of ketamine.

Continuing with the issue of the techniques used to offer a glimpse of the history of ketamine and its use in patients with mental illness, it is *de rigueur* to return to Berrios.[15,18] In studying the history of mental illnesses, Berrios suggests specifying if we are tracking the history of some of (1) the terms (in this chapter, 'dissociation' is an essential term that drives the narrative) or the historical meanderings of (2) the concepts attached to those terms or, finally, (3) the behaviors of the patients.

The task of writing a history of ketamine and its use in mental illness becomes particularly complex when the nosology and nosography of mental illnesses is also in flux, and this is illustrated, for example, by the choice of terminology for the action of ketamine as an anesthetic; I am referring to the term 'dissociative anesthesia.' Paraphrasing Farrell[19] trying to write this history against a changing psychiatric culture "[…] is akin to measuring a developing weather front with a stepladder and a yardstick."

One of the most interesting aspects of the introduction of ketamine to treat depression (and acute suicidal ideation) is the potential implications that the polymorphic nature of the altered 'states of consciousness' induced by this substance can bring not only to the terrain of the pathophysiology of depression but also to our current understanding of the pathogenesis of other disorders.

The origin of the term dissociative anesthesia

The story of how the wave of psychotomimetic effects caused by ketamine and its study leads to the term 'dissociative anesthesia' has been covered several times; however, and for the purposes of this chapter, it deserves some attention. In an often-quoted paper titled Taming the Ketamine Tiger, Edward Domino[20] described how he was asked to work with the forerunner of ketamine, phencyclidine (PCP), by his mentor, Dr. Maurice Seevers,[21] who was the Head of Pharmacology at the University of Michigan. Domino then recollects his discovery that PCP could produce 'emergency delirium' in dogs and how PCP acted as a "remarkable anesthetic" in monkeys.[20] PCP was then utilized as an anesthetic in humans by Dr. Ferdinand Greifeinstein,[22] who was the Chair of Anesthesiology at Wayne State University, Detroit, Michigan and Detroit Receiving Hospital. However, and due to the unpredictable occurrence of side effects, in particular a "state of prolonged emergency delirium",[23] PCP was voluntarily discontinued as a general anesthetic in 1965.[20] This setback did not deter the scientists at Parke-Davis; a second substance of the cyclohexylamine group, cyclohexamine (CI400), was released for use in anesthesia in 1960.[24] The presence of agitation and hallucinations in a significant proportion of women who took cyclohexamine added to the fact that only 19 of the 29 women experienced analgesia, motivating the dismissal of cyclohexamine.

Another substance synthesized by Dr. Calvin Stevens, working for Parke-Davis, known as CI-581 (ketamine), called the attention of Dr. Alex Lane, who was the Head of Clinical Pharmacology at Parke-Davis. Dr. Lane approached Edward Domino,[20] who agreed to conduct the study in humans. Domino then contacted Dr. Guenter Corssen, a professor in anesthesiology interested in intravenous anesthesia. According to Domino, the first time a human was given ketamine in an intravenous subanesthetic dose was on August 3, 1964, at the Jackson Prison in the State of Michigan.

In his 2010 paper, Domino goes on to the describe some of the remarkable effects of ketamine, including the minimal occurrence of "frank emergence delirium." Domino contacted his colleague Dr. Elliot Luby, who was a psychiatrist; Parke-Davis feared the schizophrenomimetic effects of ketamine would render it useless. The psychiatrists (in plural in the 2010 paper by Domino) working for Parke-Davis concluded that the emergence reaction of the subjects tested was similar to the emergence reaction of diethyl ether. Ketamine then received the green light.[25]

The choice of the term dissociative anesthetic is interesting. Domino describes how there was "a good deal of discussion" among Dr. Guenter Corssen, Dr. Peter Chodoff, and himself regarding the publication of the data. Corssen and Chodoff were both anesthesiologists. They initially agreed to use the term "dreaming" because it was similar to the effects of PCP. According to Domino, the Parke-Davis scientists did not like it. Domino shared his concerns with his wife, Antoinette (Toni) Domino. He told her the subjects were "disconnected" from their environment. According to Domino, Toni came up with the term dissociative anesthetic.[20] In another paper,[26] Domino recounts this story with slight variations; Toni, his wife, is the protagonist in both and, moreover, the person responsible for using the term 'dissociation.'

Domino's version was, however, in contrast to Guenter Corssen's version,[27] who gave a somewhat different reason for using the terms 'dissociation' and 'dissociative.' For Corssen the reason for choosing 'dissociative' stemmed from the actual physiologic impact of the new medication. In his paper, he describes how Domino and himself observed the way in which the visual and somatosensory stimuli traveled "unimpaired from the periphery to the primary sensory cortex, indicating that the sensory isolation occurred within the brain, presumably in the *association* area" (the italics are mine). Furthermore, electroencephalographic studies done in cats under the effect of ketamine had shown the presence of theta arousal waves in the hippocampus, with hypersynchronous delta wave burst in the area of the neocortex and in the thalamus. According to Corssen, "the electrophysiological dissociation of thalamoneocortical and limbic systems prompted us to use the term 'dissociative anesthesia'."

Incidentally, the paper by Domino, Chodoff, and Corssen became a 'citation classic'[28] in Eugene Garfield's nomenclature (i.e., 290 citations since its publication in 1965). In the comment that Domino wrote of in his 1965 paper, he did not use the term 'dissociation' but described the study subjects as "zombies who were totally disconnected from their environment." This is a far cry from the 'dissociation' chosen by his wife Toni just before its publication in 1965.

There are thus two narratives, not mutually exclusive, behind the term dissociative anesthesia. What is obviously implied is that a drug, in this case ketamine, caused anesthesia by provoking the phenomenon of dissociation; in other words, ketamine causes anesthesia by provoking dissociation. For Corssen, dissociation was merely an electrophysiologic mechanism, with an electroencephalographic correlate comprising several encephalic structures, in particular the 'association' cortex, hippocampi, and thalami; for Domino, the behavioral correlates were the 'dissociative' ones. In following Berrios'[18] methodological principles, Edward Domino appeared to focus his attention on some of the behaviors displayed by his patients, whereas Corssen centered his attention on a biological signal *sui generis*. Domino's and Corssen's views appear, at face value, to be clearly divergent. There is, however, an interesting twist to this apparent divergence: it is possible that Domino had forgotten a paper published in 1968 with Miyasaka,[29] in which he was the second author, and in which they describe the effects of ketamine on the EEG of cats. Corssen appears to have copied word by word the findings of Miyasaka and Domino in his paper of 1985.

Part of the history of ketamine consists of tracing the vicissitudes (i.e., the interactions, convergences, and divergences of behaviors, terms, concepts, and 'biological signatures' over historical periods) of the notion of dissociation as it was understood by the successive players in the field of anesthesia, neurochemistry, and later on psychiatry during the following decades.

A note on the history of dissociation

In a paper on the concept of dissociation in psychiatry, Berrios[30] deplores the lack of interest in the ontology (internal structure), epistemology (way of knowing it), and the mechanisms of dissociation. For Berrios, the scientific literature on the subject of dissociation "seem to take for granted the conceptual intelligibility and plausibility of dissociation." He then posits that "[dissociation's] very vagueness has made" it popular, in addition to having an attractive "metaphorical" appeal.

Berrios reports with his characteristic terse prose and detail that the concept of dissociation arrived in psychology and psychiatry in the 19th century from the fields of physics and chemistry. He goes on to say that there is little evidence that the choice of several terms (i.e., splitting, sejunction, disconnection, disharmony, discordance, separation, cleavage, etc.) expressed nuanced theoretic views on how dissociation was to be used as an explanation. Furthermore, for Berrios the terms are likely to be the historical expression of international needs for originality and novelty. In essence, he concludes that little is known about what he calls the operational and functional attributes of dissociation; he then enumerates several of them, including the actual orientation of the plane of cleavage (horizontal or vertical); the energy consumed by the disconnection; the mechanisms setting the components asunder; the status of dissociation as an ordinary, 'normal' psychologic mechanism that can become pathologic in certain situations; and so on.

Despite the critical comments made by Berrios in regard to "add[ing] on to a repetitive and boring clinical literature," it appears prudent to access some recent chronology of the conceptual models of dissociation to understand the gist of this section. In one of those 'boring' (Berrios's dixit) articles, Onno van der Hart and Martin Dorahy[31] write that the concept of dissociation attracted minimal attention in the 1940 and 1950s; this extended throughout the 1960s, culminating in 1969 with the publication of Charles Tart's volume Altered States of Consciousness. For van der Hart and Dorahy, the opening chapter in Tart's book, written by Arnold Ludwig,[32] was important because it brought to the fore the notion of 'Altered States of Consciousness' (ASC) in terms that, although were not analogous to 'dissociation,' were, according to them, "clearly comparable to contemporary understanding of the dissociative continuum." However, it is difficult to reconcile the broad perspective of Arnold Ludwig, who included disparate phenomena (e.g., from the effects of solitary confinement to the emergence of trance states experienced during prolonged masturbation—I have quoted literally from Ludwig's chapter), with the short description of the 'Dissociative Reaction,' described in the 1952 edition of the Diagnostic and Statistical Manual of the American Psychiatric Association,[33] which was in use when Domino, Chodoff, and Corssen published their initial findings. If anything, Ludwig's ASC would include the 'Dissociative Reaction' of *DSM-I*.

By the beginning of the 1980s, *DSM-III* "had changed everything" in the words of John Kihlstrom.[34] The shift from a psychodynamic model of mental illness to a more strictly biological one literally caught psychodynamic psychiatry unawares.[35] In the words of Stephen Hyler and Robert Spitzer, hysteria was split asunder.[36] One of the unintended consequences of *DSM-III* was the rediscovery of Pierre Janet's alternative psychodynamic model,[37] the increasing evidence of some kind of connection between trauma and dissociative phenomena, and the mushrooming of the interest in dissociative processes and phenomena.[38]

In light of the above, it is not unreasonable to think that the term 'dissociative' in 'dissociative anesthesia' had little or no connection whatsoever with 'dissociation' as it was understood after the introduction of the *DSM-II* (in the mid-1960s) and later on with the transformative effects of *DSM-III*.

From Phencyclidine and ketamine to model psychosis

In a 1959 paper, Elliot Luby wrote about the "schizophrenomimetic" effects of PCP.[23] In the 1960 and 1970s, PCP and ketamine were the focus of intense research owing to their psychotomimetic effects. Domino and Luby[39] summarized the efforts of academic and industry researchers during the 1970 and 1980s to come up with a pharmacologic model of psychosis equivalent to schizophrenia. For Lodge and Mercier,[40] those efforts were so important that pharmaceutical companies invested heavily in searching for antagonists of PCP. This is reflected in a 1979 workshop held at the annual meeting of the American College of Psychopharmacology to stimulate research in this area and in another meeting in Montpellier, France, organized by Domino in 1983; both gatherings summarized in two books research on the area of PCP and model psychosis.[41,42] Research in this area was also catalyzed by the discovery of Anis and Lodge of the selective antagonism of ketamine on the *N*-methyl-D-aspartate (NMDA) subtype of the glutamate receptor. It is worth mentioning that Anis and Lodge sent their manuscript to the journal Nature but was rejected on the grounds of 'not being of sufficient interest'.[40]

By the 1990s, ketamine was one of the tools used to model psychotic episodes similar to schizophrenia. Two papers in important journals helped consolidate the role of selective antagonists of the NMDA receptor in the genesis of schizophrenomimetic psychoses: first, the paper by Javitt and Zukin[43] and second, the paper by Krystal and his collaborators.[44]

The paper by John Krystal is historically important not only because of its intrinsic scientific value—it is a very detailed study of the effect of intravenous ketamine on 'normal' volunteers—but also because of its methodological implications. First, because the authors became aware of the euphoriant effect of ketamine, they focused on its overwhelming schizophrenomimetic effects; in this sense, this paper represents the canonization of ketamine as one of the most important causative agents of the model psychosis with conspicuous schizophrenomimetic features.

Second, the use of a tool to measure dissociation reflected the prevailing taxonomic perspective of this complex construct and, in itself, also helped define the criteria and understand dissociative disorders in what Ian Hacking has called a 'looping effect'[45] and Sierra and Berrios[46] called, in a less acrobatic way, "the definitional instability of the dissociative disorders."

The tool in question, the Clinician Administered Dissociative State Scale (in brief CADSS), has an interesting history: the authors did not validate it until years after the publication of Krystal's paper. In the validating 1998 paper,[47] the authors acknowledge that their findings "should be considered preliminary" for several reasons, including the need for more extensive testing of reliability and validity in a larger number of subjects and the need to further testing in different populations.

Third, it is necessary to clarify that the CADSS was validated in a sample of patients with posttraumatic stress disorder and dissociative symptoms. Finally, and perhaps what is more important, many of the papers that came after this one on ketamine and depression, suicide, and so on followed *grosso modo*, the methodology introduced by Krystal and his collaborators (see, for example, the paper by Pomarol-Clotet and his collaborators in 2006,[48] which represents a refinement of Krystal's aforementioned 1994 paper).

Six years after John Krystal's fundamental paper on the psychotomimetic effects of ketamine, he went on to publish another one, which became the fons et origo of the current "ketamine wave".[49] At that point, ketamine was not the 'antidepressant diva' that it has turned to be nowadays but was an important agent in the genesis of a model psychosis that was going to allow researchers to gain insight into the pathophysiology and treatment of schizophrenia.

Dumb me, dumb me!

In this section, I have selected two publications by Edward Domino, one in English[50] and another one, perhaps little known, in Spanish,[51] and the truly engrossing interview by Denomme in 2018.[26]

In his 1999 publication, Edward Domino lamented how he had "missed the boat on chlorpromazine," leading him to become involved "with new drugs in anesthesia." It is interesting that in this paper he does not mention ketamine. However, he mentions 'neuroleptoanalgesia,' which combined fentanyl and droperidol. Droperidol, a butyrophenone, is the pretext to talk about his incursion in the field of hallucinogenic substances that provoked bizarre visual experiences by mimicking the effect of serotonin in the brain. Domino reminds us of the presence of tryptamine in traces in the body and its hallucinogenic effects when administered intravenously in massive amounts. He concludes by stating, "By the 1970s, the indole hallucinogen hypothesis of schizophrenia [was] worth pursuing." Domino became aware of a dopamine and serotonin antagonist called methiothepin; he determined that methiothepin was "an excellent dimethyltryptamine (in brief, DMT) antagonist."[52] For reasons out of his control, involving a poor relationship with 'communist politicians' in the former republic of Czechoslovakia, he could not pursue research in the area of pharmacologic treatment of schizophrenia. Domino goes on to mention that Janssen Pharmaceuticals, Inc., tested, many years later, potential antipsychotics with a mixture of D2 and 5-HT2 antagonist actions that led to the development of risperidone as one of a new generation of atypical antipsychotics. He concludes, with some longing, with the following: "Shades of methiothepin again!"

In the other text selected, a chapter written in Spanish, presumably translated from English by the editors,[51] Domino shared his thoughts on the fruitful collaboration of his academic center with pharmaceutical companies and in particular with Parke-Davis (a subsidiary of the pharmaceutic company Pfizer since the year 2000). He also goes to mention research with the molecule dizocilpine (incidentally, a product of Merck & Co. Inc., patented in 1983 and also called MK-801). Dizocilpine was a sympathomimetic with anxiolytic and anticonvulsant properties; Domino observed that in epileptic patients, dizocilpine caused an improvement of their mood.[53] Domino also quoted a paper by Reimherr and collaborators[54] who administered dizocilpine to adults with attention deficit disorder. According to Domino, the most manifest effect of MK-801 was the improvement of the mood of these patients. Domino did not seem to be particularly impressed by the 'euphoriant' effect of MK-801. In fact, in an interesting digression, he mentions a graduate researcher in his laboratory who admitted to using dizocilpine recreationally and taking some of the actual MK-801 research sample; this graduate reported experiencing psychotomimetic effects identical to the ones caused by PCP. Domino goes on to comment upon the ethical and security issues raised by this anomalous event. He concludes his chapter by proposing that low concentrations of nitrous oxide could provoke schizophrenomimetic effects. It is obvious that Domino's main concern was the schizophrenomimetic effect of PCP and its congeners.

In 2018, Domino revisits his experiences with ketamine in a fascinating interview with Nicholas Denomme.[26] At this point, Domino, who was born in 1924, is 93 years old. His work with synthetic cannabinoids (which is another absorbing read in itself) enhanced his reputation of being an expert in the dependence on cannabis. In 1982 or thereabouts, he started seeing patients with illegal substance-use problems at The Lafayette Clinic (a state-funded mental health clinic in Detroit). Domino recalls how he was asked to see a young woman who was using ketamine every couple of weeks. When he asked her why she was taking ketamine, the patient told him she felt 'depressed.' Domino told the patient she could not take ketamine to get 'high' and advised her to get her psychiatrist to taper off ketamine. Domino, as was his wont, shared his concerns with his wife: "This is a crazy drug, you get 'high' and feel good for a period after. This will never work. Dumb me, dumb me!" Domino then adds that "[...] the concept of getting a high and then having an antidepressant effect made no sense" to him. He lamented not studying this effect further ("Dumb me, dumb me!" – sic). In a final twist, and according to Domino, the idea was lost until a friend of his, John Krystal and his collaborators, showed that ketamine had an antidepressant effect.[49] He concludes the interview by wondering what would have happened if only he had listened to the patients.

Final reflections

First, we have seen how the term dissociative anesthesia had no or very little relationship with what the 'Bible of Psychiatry'[55] called 'dissociative disorders' or states. Moreover, we have also learned that the scale utilized to grade some of the emergent behaviors and symptoms provoked by ketamine was not 'validated' until 1998, that is, 4 years after an important paper describing some of the effects of ketamine in normal volunteers—this is an interesting anachronism. Third, the prevailing role of ketamine (and its analogues) was as inducer of a 'model psychotic' state. This 'model' was very present throughout Domino's intellectual and scientific career and is mentioned in several of his papers well into the late 1990s; it is historically valid to say that the scientific community, inspired by a hard-core 'pharmacocentric' perspective,[56,57] followed that epistemological format with some of its negative unintended consequences. Fourth, serendipity[58] and not the frequently repeated (and clearly exhausted) 'paradigm shift' contributed to the belief that ketamine had a mood elevating effect in some patients with affective disorders. In this sense, it is worth giving Vazquez's paper a read,[59] who has provided a clinical perspective to Baldessarini's somewhat existential overview of psychopharmacology.

The history of ketamine is interesting for a number of reasons. Part of them have been mentioned in the preceding paragraphs. Ultimately, it is interesting because psychiatry, in the words of Germán Berrios,[60] is not a contemplative activity and we need new ways to look into the suffering and emotional cost of the Hippocratic triangle of depression, patient, and clinician, which is trapped in the current taxonomic and therapeutic dead ends.

References

1. Shorter E, Tyrer P. Separation of anxiety and depressive disorders: blind alley in psychopharmacology and classification of disease. *Br Med J*. 2003;327(7407):158–160.
2. Tyrer P, Tyrer H, Guo B. The general neurotic syndrome: a Re-evaluation. *Psychother Psychosom*. 2016;85(4):193–197.
3. Hyman SE. Psychiatric drug development: diagnosing a crisis. *Cerebrum*. 2013;2013:5.
4. Krystal JH, Abdallah CG, Sanacora G, Charney DS, Duman RS. Ketamine: a paradigm shift for depression research and treatment. *Neuron*. 2019;101(5):774–778.
5. Kuhn TS. *The Structure of Scientific Revolutions*. University of Chicago Press; 2012.
6. Horwitz AV. *Creating Mental Illness*. University of Chicago Press; 2002.
7. Shorter E. The history of DSM. In: *Making the DSM-5*. Springer; 2013:3–19.
8. Hartogsohn I. Constructing drug effects: a history of set and setting. *Drug Sci Pol Law*. 2017;3.
9. Mischel NA, Kritzer MD, Patkar AA, Masand PS, Szabo ST. Updates on preclinical and translational neuroscience of mood disorders: a brief historical focus on ketamine for the clinician. *J Clin Psychopharmacol*. 2019;39(6):665–672.
10. Wei Y, Chang L, Hashimoto K. A historical review of antidepressant effects of ketamine and its enantiomers. *Pharmacol Biochem Behav*. 2020;190:172870.
11. Chang LC, Rajagopalan S, Mathew SJ. The history of ketamine use and its clinical indications. In: *Ketamine for Treatment-Resistant Depression*. 2016:1–12.
12. Mion G. History of anaesthesia: the ketamine story – past, present and future. *Eur J Anaesthesiol*. 2017;34(9):571–575.
13. Shephard B. Risk factors and PTSD: a historian's perspective. In: Rosen GM, ed. *Posttraumatic Stress Disorder*. England: John Wiley & Sons, Ltd.; 2004:39–61.
14. Morris E. *The Ashtray:(Or the Man Who Denied Reality)*. University of Chicago Press; 2018.
15. Berrios GE. *Historical Research in Research Methods in Psychiatry*. RCPsych Publications; 2006.
16. Wallace E. *Psychiatry and its Nosology: A Historico-Philosophical Overview. Philosophical Perspectives on Psychiatric Diagnostic Classification*. 1994:16–86.
17. Duffin J. *A Hippocratic Triangle: History, Clinician-Historians, and Future Doctors*. 2004.
18. Berrios GE. *The History of Mental Symptoms: Descriptive Psychopathology since the Nineteenth Century*. Cambridge University Press; 1996.
19. Farrell K. *Post-traumatic Culture: Injury and Interpretation in the Nineties*. JHU Press; 1998.
20. Domino EF. Taming the ketamine tiger. 1965. *Anesthesiology*. 2010;113(3):678–684.
21. Shorter E. *An Oral History of Neuropsychopharmacology*. Vol. 1. USA: ACNP; 2011.
22. Meyer JS, Greifenstein F, Devault M. A new drug causing symptoms of sensory deprivation. *J Nerv Ment Dis*. 1959;129(1):54–61.
23. Luby ED, Cohen BD, Rosenbaum G, Gottlieb JS, Kelley R. Study of a new schizophrenomimetic drug; sernyl. *AMA Arch Neurol Psychiatry*. 1959;81(3):363–369.
24. Pender JW. Dissociative anesthesia. *J Am Med Assoc*. 1971;215(7):1126–1130.
25. Domino EF, Chodoff P, Corssen G. Pharmacologic effects of CI-581, a new dissociative anesthetic, IN MAN. *Clin Pharmacol Ther*. 1965;6:279–291.
26. Denomme NBS. The Domino effect: ed domino's early studies of psychoactive drugs. *J Psychoactive Drug*. 2018;50(4):298–305.

27. Corssen G. Dissociative anesthesia. In: *Anaesthesia*. Springer; 1985:92–94.
28. Domino EF. This Week's Citation Classic – pharmacologic effects of CI-581, a new dissociative anesthetic, in man. *Curr Contents*. 1984:306.
29. Miyasaka M, Domino EF. Neural mechanisms of ketamine-induced anesthesia. *Int J Neuropharmacol*. 1968;7(6):557–573.
30. Berrios GE. The concept of 'dissociation' in psychiatry. *Rivista Sperimentale de Freniatria*. 2018;142:29–50.
31. van der Hart O, Dorahy MJ. *History of the Concept of Dissociation*. 2009.
32. Ludwig AM. *Altered States of Consciousness*. John Wiley & Sons, Ltd.; 1969.
33. Association AP. *Diagnostic and Statistical Manual: Mental Disorders*. American Psychiatric Association; 1952.
34. Kihlstrom JF. *One Hundred Years of Hysteria*. Dissociation: Clinical and Theoretical Perspectives. 1994:365–394.
35. Wilson M. DSM-III and the transformation of American psychiatry: a history. *Am J Psychiatry*. 1993;150(3):399–410.
36. Hyler SE, Spitzer RL. Hysteria split asunder. *Am J Psychiatry*. 1978;135(12): 1500–1504.
37. Ellenberger HF. *The Discovery of the Unconscious: The History and Evolution of Dynamic Psychiatry*. Vol. 1. New York: Basic Books; 1970.
38. Cardeña E. *The Domain of Dissociation. Dissociation: Clinical and Theoretical Perspectives*. 1994:15–31.
39. Domino EF, Luby ED. Phencyclidine/schizophrenia: one view toward the past, the other to the future. *Schizophr Bull*. 2012;38(5):914–919.
40. Lodge D, Mercier MS. Ketamine and phencyclidine: the good, the bad and the unexpected. *Br J Pharmacol*. 2015;172(17):4254–4276.
41. Domino EF. *PCP (Phencyclidine): Historical and Current Perspectives*. Npp Books; 1981.
42. Kamenka J-M, Domino EF, Geneste P. In: *Phencyclidine and Related Arylcyclohexylamines: Present and Future Applications: Proceedings of the Joint French-US Seminar on the Chemistry, Pharmacology, Present and Future of the Therapeutic Applications and Drug Abuse Aspects of Arylcyclohexylamines Held in La Grande Motte (Montpellier), France, September 20–24, 1982*. Npp Books; 1983.
43. Javitt DC, Zukin SR. Recent advances in the phencyclidine model of schizophrenia. *Am J Psychiatry*. 1991;148(10):1301–1308.
44. Krystal JH, Karper LP, Seibyl JP, et al. Subanesthetic effects of the noncompetitive NMDA antagonist, ketamine, in humans. Psychotomimetic, perceptual, cognitive, and neuroendocrine responses. *Arch Gen Psychiatry*. 1994;51(3):199–214.
45. Tsou JY. Hacking on the looping effects of psychiatric classifications: what is an interactive and indifferent kind? *Int Stud Philos Sci*. 2007;21(3):329–344.
46. Sierra M, Berrios GE. Depersonalization: a conceptual history. *Hist Psychiatr*. 1997; 8(30):213–229.
47. Bremner JD, Krystal JH, Putnam FW, et al. Measurement of dissociative states with the clinician-administered dissociative states scale (CADSS). *J Trauma Stress*. 1998;11(1): 125–136.
48. Pomarol-Clotet E, Honey G, Murray G, et al. Psychological effects of ketamine in healthy volunteers: phenomenological study. *Br J Psychiatry*. 2006;189(2):173–179.
49. Berman RM, Cappiello A, Anand A, et al. Antidepressant effects of ketamine in depressed patients. *Biol Psychiatry*. 2000;47(4):351–354.

50. Domino EF. History of modern psychopharmacology: a personal view with an emphasis on antidepressants. *Psychosom Med.* 1999;61(5):591–598.
51. Domino EF. *La colaboración entre las instituciones académica y la industria farmacéutic en el desarrollo de la psicofarmacología: el ejemplo de los antagonistas NMDA fenciclidina, ketamina y dizocilpina.* Vol. 3. Médica Panamericana; 2007.
52. Moore R, Demetriou S, Domino E. Effects of iproniazid, chlorpromazine and methiothepin on DMT-induced changes in body temperature, pupillary dilatation, blood pressure and EEG in the rabbit. *Arch Int Pharmacodyn Ther.* 1975;213(1):64–72.
53. Troupin ASMJ, Cheng F. *MK-801.* Vol. 4. John Libbey Eurotext; 1986.
54. Reimherr FW, Wood DR, Wender P. The use of MK-801, a novel sympathomimetic, in adults with attention deficit disorder, residual type. *Psychopharmacol Bull.* 1986;22(1): 237.
55. Wilson HS, Skodol A. Special report: DSM-IV: overview and examination of major changes. *Arch Psychiatr Nurs.* 1994;8(6):340–347.
56. Baldessarini RJ. *American Biological Psychiatry and Psychopharmacology, 1944–1994. American Psychiatry after World War II: 1944–1994.* 2000:371–412.
57. Baldessarini RJ. The impact of psychopharmacology on contemporary psychiatry. *Can J Psychiatry.* 2014;59(8):401–405.
58. Yaqub O. Serendipity: towards a taxonomy and a theory. *Res Pol.* 2018;47(1):169–179.
59. Vazquez GH. The impact of psychopharmacology on contemporary clinical psychiatry. *Can J Psychiatry.* 2014;59(8):412–416.
60. Berrios GE. Classifications in psychiatry: a conceptual history. *Aust N Z J Psychiatr.* 1999;33(2):145–160.

CHAPTER

2 Ketamine's potential mechanism of action for rapid antidepressive effects – a focus on neuroplasticity

Melody J.Y. Kang, BScH

Master of Science, Centre for Neuroscience Studies, Queen's University, Kingston, ON, Canada

Background

Major depressive disorder (MDD) affects an estimate of 264 million people, and it is currently the leading cause of disability internationally.[1] Current antidepressants take approximately two to four weeks to take into effect, and up to 46% of patients do not respond adequately.[2] This delay in efficacy results in a window of time when patients experience prolonged suffering and even worsening of symptoms, giving rise to a dire need for novel treatments. Recent efforts in the development of innovative antidepressants has generated interest and promoted research surrounding subanesthetic ketamine. Ketamine is an *N*-methyl-D-aspartate receptor (NMDAR) antagonist in the glutamatergic system and has been used for decades as an anesthetic at high doses (Fig. 2.1). More recently, it has gathered widespread attention due to its rapid antidepressant effects at lower doses (0.5 mg/kg), and many studies have shown significant efficacy in patients who have not responded to classic monoaminergic antidepressants. Response rates for ketamine are approximately 60% –70%, with some patients reporting improvements in mood as soon as 40 minutes

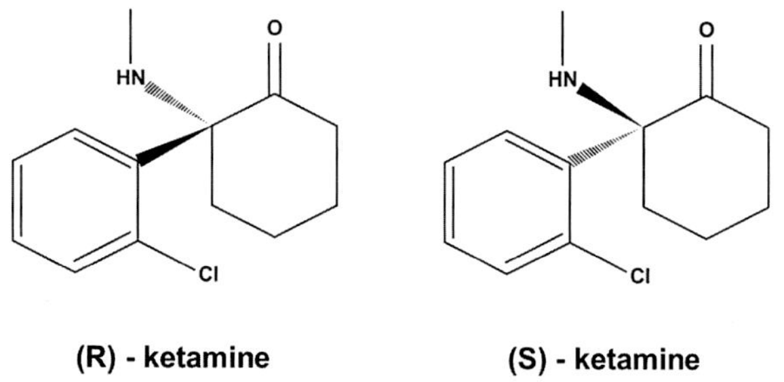

(R) - ketamine

(S) - ketamine

FIGURE 2.1

Two stereoisomers of ketamine.

Ketamine for Treatment-Resistant Depression. https://doi.org/10.1016/B978-0-12-821033-8.00002-2

after infusion.[3–8] Although it is clear that ketamine is efficacious for many, it has the potential to be used as a drug of abuse with potent side effects, such as dissociation.[9,10] This restricts its ability to be prescribed as a widespread antidepressant. This indicates a crucial need for effective and novel treatments that preserve and mimic ketamine's rapid effects while lacking its side effects; however, ketamine's mechanism of action is currently unknown.

One neurobiological theory for depression is the neuroplasticity hypothesis, which suggests that deficits in neuroplasticity may be responsible for the pathology of MDD. Although there is no formal definition, many define neuroplasticity as the brain's ability to restructure itself upon internal or external stimuli.[11–15] MDD has been robustly linked to atypical neuroplasticity through preclinical and clinical studies, which show decreased levels of neuroplastic markers, reduced number of critical neuronal structures such as dendrites or synapses, or decreased brain volumes.[16,17] In addition, traditional antidepressants have shown to restore abnormalities of neuroplasticity in animal models of depression.[18–20] Through these developments, neuroplasticity has come to be a potential target in developing novel treatment options for MDD.

Our treatment of interest, ketamine, has shown to manipulate many domains involved in neuroplasticity. Current literature suggests that the neuroplastic mechanisms involved in ketamine's rapid (minutes to hours) antidepressant effect consist largely of markers of molecular neuroplasticity and diverge from those that contribute to its potential sustained (days to weeks) effects. However, the relationship among the pathology of depression, ketamine's ability to restore molecular neuroplasticity, and the resulting rapid antidepressant effects remains unclear. This chapter aims to assert that ketamine's mechanism of action heavily involves altering and restoring molecular neuroplastic mechanisms in the brain to induce its rapid antidepressant effect. The chapter summarizes the current field of literature of ketamine's impact on molecular neuroplasticity and briefly discusses other mechanisms that likely contribute to its effect.

Ketamine restores molecular neuroplastic molecules to induce rapid antidepressant effects

Multiple molecular changes may work either cohesively or consequentially like a domino effect in order for ketamine to produce the rapid change in behavior and mood. Molecules of interest that have been associated with ketamine include glutamate/glutamine, α-amino-3-hydroxy-5-methyl-4-isoxazolepropionic acid receptors (AMPARs), mechanistic target of rapamycin (mTOR), brain-derived neurotrophic factor (BDNF), VGF, eukaryotic elongation factor 2 kinase (eEF2K), p70S6K, glycogen synthase kinase 3 (GSK-3), insulin-like growth factor 2 (IGF-2), MAPK/Erk, and microRNAs. Changes in activation levels or quantities are shown in preclinical studies using cultured neurons and animal models and in clinical studies involving patients with depression. Together they suggest that improvements in mood or antidepressant-like behavioral effects observed after ketamine administration are strongly correlated with enhanced molecular neuroplasticity in the brain.

Glutamate, glutamine, and γ-aminobutyric acid availability and cycling

Glutamate is the main excitatory neurotransmitter in the brain and is normally used to consolidate learning and memory. It has wide-ranging functionality in over half of all synapses. Glutamate must be synthesized in our nerve cells, as it cannot cross the blood-brain barrier, and the most prevalent method of synthesis is with the use of its precursor, glutamine, through the glutamate-glutamine cycle. Conversely, γ-aminobutyric acid (GABA) is the main inhibitory neurotransmitter in the brain and it regulates excitatory signals or action potentials in the brain. Multiple articles have postulated that what is responsible for the *rapid* antidepressant effect of ketamine is its unique ability to induce a "glutamate burst" that occurs after administration at the synapse of pyramidal neurons in the prefrontal cortex.[21–24] This glutamate burst is not observed with other traditional antidepressants that take weeks to come into effect. However, recent preclinical and clinical studies show that there may be more to this mechanism regarding the levels and activity of glutamate in other brain structures, in addition to the cortex.

Acutely, after ketamine administration, we observe increased activity and cycling of glutamate and glutamine in the prefrontal cortex of humans.[25] In addition, ketamine induces levels of glutamate, glutamine, and GABA to increase in the prefrontal cortex in healthy participants,[26] as well as in patients with major depression.[27] Notably, this increase peaks at 26 minutes into the administration, which may underlie psychomimetic changes rather than representing the mechanism behind the antidepressant effect.[27] GABA, the inhibitory neurotransmitter, is subsequently shown to decrease in concentration over time, i.e., 3–4 hours after infusion in healthy participants.[28] This trend of fluctuating glutamate, glutamine, and GABA levels is observed in preclinical studies as well, particularly in the prefrontal cortex and hippocampus of rodents.[29–31] Ketamine restores levels of glutamate transporters 2 and 3 in the hippocampus of stressed mice, which may suggest increased glutamate-glutamine cycling.[32] Continuing the pattern of decreased GABA levels overtime, chronic ketamine administration results in overall decreased levels of GABA and GABA-synthesizing enzymes, GAD67 and GAD65, in the cortex, cerebellum, and striatum in rats.[33] We theorize that ketamine may leave the cortex in a more "excitatory" state after infusion.

Interestingly, in the anterior cingulate cortex (ACC), we do not observe similar patterns. GABA level is significantly increased 24 hours after ketamine administration in patients with major depression, possibly from glutamate to GABA synthesis.[34] In healthy controls, glutamine level is shown to increase in the ACC after 24 hours, but glutamate levels remain unchanged.[35,36] These alterations in glutamatergic transmission in varying brain regions may directly contribute to behavioral changes or trigger a signaling cascade by binding with receptors on the postsynaptic terminal that results in rapid relief of depressive symptoms.

Enhanced activation of the α-amino-3-hydroxy-5-methyl-4-isoxazolepropionic acid receptor

Glutamate can bind to three different identified ionotropic receptors, one of which is AMPAR. This is composed of four subunits: GluA1, GluA2, GluA3, and GluA4.[37,38] Located on the postsynaptic membrane of neurons, the AMPAR functions to mediate synaptic transmission and plasticity.[37,39] Unsurprisingly, increased levels of glutamate after ketamine administration enhance AMPAR activity, first shown in 2008.[40] Numerous reports since then have provided strong evidence to outline AMPAR involvement in ketamine's mechanism of action. In rodent models of depression, ketamine's rapid antidepressant-like behavioral effects are abolished upon administration of NBQX, an AMPAR antagonist.[40–42] Conversely, AMPAR agonists enhance ketamine's antidepressant-like behavioral effects when combined with a dose of ketamine below the typical threshold of response.[43] In fact, even when AMPAR agonists are administered alone, rats still exhibit antidepressant-like behaviors.[41]

In addition to behavioral alterations, we observe AMPAR involvement in changes to levels of protein and messenger RNA (mRNA). Ketamine administration increases membrane levels of GluA1 in mice hippocampi.[44,45] AMPAR agonists increase levels of downstream proteins previously associated with ketamine and neuroplasticity, such as BDNF and mTOR signaling proteins, in vivo.[41] Ketamine-induced increases in these protein levels are inhibited by AMPAR antagonists in the hippocampus and prefrontal cortex of rodents.[41,46,47] Reciprocally, inhibiting mTOR signaling reduces AMPAR activation.[48] The levels of AMPAR/NMDAR ratio transiently increase after ketamine exposure in the hippocampus of a rat model of anxiety.[49] Cultured neurons from the dorsal raphe nucleus reveal ketamine-induced increases in the mRNA of AMPAR subunits (GluA1, GluA2, and GluA4).[50] The phosphorylation status of the GluA1 subunit and its role in ketamine's mechanism are unclear, as both increased[51] and decreased[40] phosphorylation is reported to be induced by ketamine in the hippocampus of rodents.

Increased mechanistic target of rapamycin signaling pathway activation

A robust amount of evidence supports the high likelihood of AMPAR involvement in ketamine's mechanism of action. However, it remains unclear which specific steps contribute and are integral to the process of inducing rapid antidepressant effects. The activation of AMPAR is followed by a cascade of signaling molecules that promote neuroplasticity, one of which is through mTOR. This protein kinase is a master regulator of cell growth that affects many upstream and downstream cellular processes and executes survival, proliferation, and maintenance of cells.[52,53] The exposure to ketamine induces an increase in the amount of proteins involved in the functions of mTOR, such as p4E-BP, p70S6K, Erk, Akt, postsynaptic density protein 95 (PSD-95), and GluA1, in the prefrontal cortex of a rodent model of depression.[54] The increases in protein levels are not observed if ketamine is combined with rapamycin, the mTOR inhibitor.[54,55]

Mimicking ketamine's pharmacology results in the manipulation of mTOR. Deletion of NR2B on interneurons, a subunit of NMDARs, simulates ketamine's antagonism upon binding. This increases levels of mTOR and shows similar behavioral antidepressant-like effects in mice.[56,57] Activation of NMDARs also blocks ketamine-induced increases in "Rheb," a G-protein that normally participates in the activation of mTOR.[58] These results support AMPAR activation and NMDAR antagonism involvement in mTOR and consequently the mechanism of action of ketamine.

Brain-derived neurotrophic factor

mTOR activation produces a neurotrophin called BDNF. BDNF, which has been highly researched in mood and antidepressant treatments, is a marker of neuroplasticity that promotes neuronal growth and survival.[59,60] Upon production, it is released into the synaptic space from the postsynaptic neuron and stimulates its receptor tropomyosin receptor kinase B (TrkB) on the same postsynaptic terminal, thus further enhancing mTOR activation and resulting in a positive feedback loop.[61,62]

The implication of BDNF-TrkB signaling in ketamine's actions is strongly supported by numerous in vivo preclinical studies. For example, upon administration of ketamine, acutely increased levels of BDNF are observed in the rat hippocampus and amygdala.[63–65] Ketamine restores levels of previously reduced BDNF in the hippocampus and prefrontal cortex of a rodent model of depression.[65–67] In animal models, when the actions of BDNF are artificially blocked with a BDNF-neutralizing antibody, with TrkB inhibitor, or by knocking out BDNF/TrkB genes, ketamine's behavioral antidepressant-like effects are inhibited.[68–71] Primary neuronal cell cultures exposed to ketamine induce rapid release of BDNF within 15 minutes, supporting the involvement of this signaling pathway in ketamine's rapid effects.[46]

The mechanism of how ketamine increases BDNF protein levels is unclear. BDNF's mRNA levels are not affected in mice models of depression after ketamine administration, suggesting that the rate of translation, not transcription, may be increased after ketamine administration.[69] In addition, ketamine may be stimulating a larger amount of BDNF precursor protein (proBDNF) to be cleaved to produce mature BDNF (mBDNF).[72] Ketamine results in increased amounts of mBDNF and a reduced ratio of proBDNF/mBDNF in the hippocampus of stressed rats.[72] When this cleavage is inhibited with a tissue plasminogen activator inhibitor, rats exhibit increased depression-like behavioral symptoms.[72]

The increase in BDNF levels we observe is acute and transient. Rats demonstrate the highest increase in BDNF levels in the prefrontal cortex, amygdala, and hippocampus when killed immediately after ketamine administration, rather than after 1 or 6-hours.[73–75] Chronic ketamine administration for 14 days resulted in unaffected BDNF protein levels in the rat hippocampus, although behavioral antidepressant-like effects were intact.[69,75,76] These results suggest that homeostatic mechanisms may clear BDNF levels overtime, stressing the likelihood of BDNF's role in rapid rather than sustained behavioral effects.[73]

The translational nature of the results from these preclinical studies should be analyzed with caution. All evidence discussed so far have used brain tissue to quantify levels of BDNF, while clinical studies will only be able to utilize peripheral protein levels. To concur, increased levels of plasma BDNF have zero correlation to BDNF levels in the rat prefrontal cortex and hippocampus.[77] Expanding on this limitation, current literature regarding BDNF levels in humans are contradictory. Increased serum/plasma levels of BDNF that correlated to mood improvements[78–80] and unchanged levels of BDNF despite mood improvements[81] have both been reported, 230–240 minutes after ketamine administration. These discrepancies in results lead to challenges in determining the specific role of BDNF in ketamine's rapid antidepressant mechanism of action.

VGF

One of the peptides regulated by BDNF is called VGF (nonacronymic). VGF is upregulated by exercise and downregulated by stress in animal models of depression,[82,83] unveiling its possible involvement in antidepressant-like effects. Upon ketamine administration, reduced levels of VGF in stressed mice are restored.[84] Inhibition of this peptide in the prefrontal cortex blocks ketamine-induced mTOR signaling and antidepressant-like behaviors in mice.[84] When VGF is knocked out, mice are more susceptible to stress paradigms (less resilient) and ketamine's behavioral improvements are reduced.[84] When VGF is artificially overexpressed, the behavioral deficiencies caused by chronic restraint stress are shown to be prevented in mice.[84] Although these results need to be reproduced, VGF seems like a promising biomarker in ketamine's mechanism of action.

Eukaryotic elongation factor 2

The protein, eukaryotic elongation factor 2 (eEF2), and eEF2K work together downstream of mTOR. eEF2K is a $Ca^{2+/}$calmodulin-dependent serine/threonine kinase that is integral for the regulation of protein translation.[85] Under normal conditions of the NMDAR, eEF2 is phosphorylated by eEF2K to make p-eEF2 and pause protein translation. It works by regulating ribosome translocation to regulate protein synthesis.[86] Upon antagonism of the NMDAR, eEF2K is inhibited and eEF2 is no longer phosphorylated resulting in increased protein synthesis. This is how we presume ketamine's function may involve eEF2. Exposure to ketamine results in decreased levels of p-eEF2 in the mice hippocampus.[69] Inhibiting eEF2K directly results in increased levels of the BDNF protein.[69] Inhibition of eEF2K in BDNF knockout mice shows no antidepressant-like behavioral effects, suggesting the importance of the relationship between eEF2K and BDNF in ketamine's mechanism of action.[69] This function, however, may depend on each brain structure, as the prefrontal cortex shows increased levels of phosphorylated eEF2K in mice.[87] The significance of this phosphorylation remains unclear, as the site of phosphorylation can dictate both increased and decreased activation of eEF2K.[85]

p70 ribosomal S6 kinase

Downstream to mTOR is a serine/threonine kinase called p70 ribosomal S6 kinase (p70S6K), which has been shown to promote protein synthesis and proliferation. p70S6K has been studied in all different types of tissues, with a heavy focus on its interrelationship with mTOR. Ketamine induces increased levels of this enzyme in in vitro primary neuronal cultures, as well as in in vivo studies including medial prefrontal cortex of rats and the nucleus accumbens, ventral tegmental area, substantia nigra, hippocampus, and basolateral amygdala of mice.[46,47,58,88] Ketamine-induced effects were blocked by the mTOR inhibitor, rapamycin.[47,58]

Glycogen synthase kinase 3

Another enzyme involved in the mTOR signaling pathway is GSK-3, which contributes to the regulation of the cell cycle, specifically proliferation and apoptosis.[89] Inhibition of this enzyme is robustly associated with ketamine's mechanism of action. For example, animal models of depression show that this inhibition is a prerequisite for ketamine's behavioral antidepressant-like effects, as constitutively active GSK-3 results in resistance to ketamine.[90,91] Doses below the threshold of response of ketamine can work synergistically with lithium, a GSK-3 inhibitor, to augment ketamine's effects and induce responses equivalent to those of higher doses of ketamine.[65] In fact, lithium alone shows similar behavioral effects in mice as ketamine.[90] Supporting the contribution of this kinase to ketamine's mechanism, the inactive form of GSK-3, called GSK-3β, increases in concentration in the rat prefrontal cortex after ketamine adminisatration.[92] Knocking out GSK-3 inhibits ketamine's effects of increasing hippocampal membrane levels of GluA1, as well as increased levels of BDNF, in mice.[44]

A potential mechanism of how ketamine interacts with GSK-3 is thought to be via the PSD-95. PSD-95 is a potent regulator of synaptic strength involved in AMPAR trafficking.[93] GSK-3 targets phosphorylated PSD-95, which promotes GluA1 internalization.[44] Ketamine exposure decreases hippocampal levels of phosphorylated PSD-95, which results in increased membrane levels of GluA1, available to be activated.[44] Ketamine also normalizes levels of PSD-95, previously downregulated in the hippocampus of rat models of depression.[94]

A protein upstream of GSK-3 called IGF-2 is another substrate thought to contribute to ketamine's mechanism of action. Mice resilient to depressive paradigms show higher levels of IGF-2 in the hippocampus.[91] IGF-2 is upregulated in the mouse hippocampus with ketamine, which requires the inhibition of GSK-3.[91] Knocking out IGF-2 reduces ketamine's behavioral antidepressant-like effects.[91]

Limitations to mechanistic target of rapamycin involvement in ketamine's mechanism of action

The current field of literature robustly supports involvement of mTOR and its accessory proteins in ketamine's mechanism of action. However, there are studies that highlight the likelihood of other signaling cascades working in conjunction to induce these effects. For example, although female rats display greater behavioral sensitivity to lower doses of ketamine than male rats, they do not demonstrate an increased level of mTOR phosphorylation, or decreased levels of eEF2K.[95] In fact, decreased phosphorylation levels of mTOR are shown in the prefrontal cortex of female mice after ketamine administration.[96] Increased sensitivity to ketamine is not observed in ovariectomized female rats but this is restored upon introduction to artificial estrogen and progesterone.[95] Another marker of plasticity, acetylation of α-tubulin, which represents increased stabilization of microtubules, only increases in female not male rats after ketamine administration.[97] These results together suggest that mTOR activation may be highly sex-dependent in animal models. These differential responses may indicate completely dissimilar mechanisms of action, or varying metabolism of ketamine. Ketamine likely does not function via mTOR exclusively and operates in part through gonadal hormones to induce rapid behavioral effects.

In addition, many preclinical studies being discussed utilize varying models of depression in rodents, introducing another variable. As the diverse models may represent different phenotypes of depression, they may not homogeneously represent one illness but multiple illnesses that manifest similar behavioral patterns. For example, when a model of *resistant* depression was induced, mTOR levels were reduced after ketamine exposure; however, antidepressant-like behavioral effects were observed.[98] These outcomes are contrary to most reports that have associated increased mTOR activation with improved behavioral effects.

Lastly, a recent clinical study shows oral rapamycin (mTOR inhibitor) prolonging ketamine's antidepressive effects at the 2-week timepoint.[99] However, rapamycin is not associated with the acute antidepressive effect.[99] This is conflicting with most results from preclinical studies that show ketamine's effects being inhibited when rapamycin is applied. It may be due to a lack of blood-brain barrier penetration of oral rapamycin, or peripheral reactions that may impact central pathways. This raises questions about the role of mTOR in acute or sustained antidepressive effects of ketamine and the extent of translation of the results collected from preclinical studies to patients with depression. These limitations together place the current field of evidence into perspective and highlights the need for further research of mTOR's role in ketamine's mechanism of action.

Inhibition of MAPK/ErK

MAPK/Erk is an important signaling pathway involved in neural plasticity that carries signals from the surface of the cell down to the nucleus. This pathway also regulates CREB, a protein integral in neuronal cell survival. Ketamine manipulates this pathway in similar ways that a MAPK inhibitor does. For example, exposure to ketamine and exposure to an MAPK inhibitor both increase levels of phosphorylated CREB in cultured neurons, as well as the prefrontal cortex of rats, whereas levels decrease in the hippocampus.[100,101] Infusion of both an MAPK inhibitor and ketamine results in an additive effect, augmenting these molecular changes even further.[101] A protein upstream of MAPK/Erk called vascular endothelial growth factor (VEGF) is an extracellular growth factor. When VEGF or its receptor is selectively deleted from neuronal cells, or a VEGF neutralizing antibody is infused in rodents, ketamine's synaptogenic or behavioral effects are inhibited.[102] Infusion of VEGF into the medial prefrontal cortex of rodents results in rapid antidepressant-like behavioral effects.[102] However, in the rat hippocampus, the level of VEGF is reported to decrease 2 hours after ketamine administration.[103] These reports suggest MAPK/Erk and VEGF interactions with ketamine to be integral in its mechanism of action, although variable upon brain region.

microRNA expression

microRNAs are small noncoding molecules that aid in protein synthesis. microRNAs from the miR-29 family are important in regulating neuropathologic processes and neuroplasticity. One of its potential targets is metabotropic glutamate receptor 4 (GRM4), which regulates neurotransmitters (i.e., glutamate, dopamine, etc.). A rat model of depression shows downregulated miR-29 and increased GRM4, which are normalized with ketamine, or even preventable with pretreatment of ketamine.[104] Restored levels of miR-29 are associated with improved behavioral effects.[104] microRNA from the miR-206 family, which is shown to critically regulate BDNF protein synthesis, is downregulated in ketamine-treated rodents.[105] Overexpression of miR-206 is shown to block ketamine-induced increase of BDNF levels in primary hippocampal pyramidal cultured neurons, suggesting that the increased BDNF levels observed after ketamine administration may occur in part via this pathway.[105]

Divergence of mechanisms for ketamine's rapid effects versus sustained effects

Numerous reports have shown enhanced structural neuroplasticity in animal models of depression following ketamine exposure. The aforementioned signaling cascades triggered by ketamine cumulatively function together to bring about increased rates of protein synthesis. This is observed in the form of synaptogenesis and spinogenesis, which are processes normally integral for learning and memory. Ketamine has shown to restore spinal density in specific locations in the rodent brain.[106–110] In addition, the formation of these new spines is associated with antidepressant-like behaviors; however, they are unlikely to contribute to ketamine's *rapid* effects. Dendritic spines only have functional synapses if sustained for 4 days, and therefore their contribution to rapid improvements may be unlikely.[111] This was further supported by Moda-Sava et al.[110] who showed rapid antidepressant-like behavioral improvements in mice being precursory to the observed spinogenesis. In fact, the generation of these structures was associated with behavioral antidepressant-like effects 2–7 days after ketamine administration, instead of rapid behavioral changes observed immediately after ketamine exposure. Furthermore, modulation and restoration of structures implicated in synaptic machinery have previously been linked to studies of monoaminergic antidepressants after chronic treatment.[14,112,113] The structural neuroplasticity demonstrated by both ketamine and monoaminergic agents is unlikely responsible for the rapidity of ketamine. These observations together support the possibility that separate mechanisms are at play to produce rapid and sustained antidepressant effects (Fig. 2.2).

Limitations to the molecular neuroplasticity mechanism and recent findings

Although ketamine's rapid effects have strong correlations with molecular neuroplasticity, it is improbable that these mechanisms function in isolation and are solely responsible for inducing rapid antidepressant effects. In fact, many other mechanisms, both related and unrelated to neuroplasticity, have shown to be manipulated after ketamine exposure and are associated with behavioral improvements.

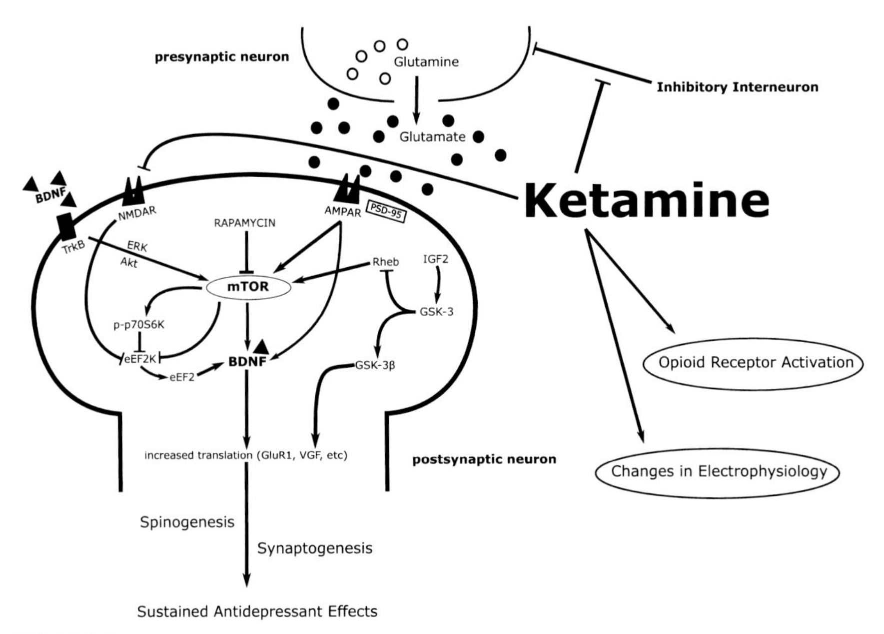

FIGURE 2.2

Representation of molecular signaling cascades that occur after NMDAR antagonism upon ketamine exposure. AMPAR, α-amino-3-hydroxy-5-methyl-4-isoxazolepropionic acid receptor; *BDNF,* brain-derived neurotrophic factor; *eEF2,* eukaryotic elongation factor 2; *eEF2K,* eukaryotic elongation factor 2 kinase; *GSK-3,* glycogen synthase kinase 3; *IGF-2,* insulinlike growth factor 2; mTOR, mechanistic target of rapamycin; *NMDAR,* N-methyl-d-aspartate receptor; *p70S6K,* p70 ribosomal S6 kinase; *PSD-95,* postsynaptic density protein 95; *TrkB,* tropomyosin receptor kinase B.

The previously mentioned glutamate burst may result from ketamine preferentially binding to NMDARs on GABA-mediated (GABAergic) interneurons, resulting in the disinhibition of pyramidal neurons.[21–24,114] The rapid effects may be due to the effects of this preferential binding, and the resulting action potential and synaptic potentiation that occur.[24,114] This mechanism of action is supported by a clinical study that showed ketamine restoring functional connectivity in the prefrontal cortex of patients with MDD,[115] which was inhibited by GABAergic interneuron agonists, benzodiazepines.[116]

Changes in electrophysiology are observed through increased excitability and neuronal potentiation. In patients with treatment-resistant depression, increases in neural responses in the right caudate were associated with improvements in mood after ketamine administration.[117] Alterations in gamma oscillations, important in learning and memory, are observed as well. Ketamine generates a sustained increase of gamma oscillations in the prefrontal cortex of rodents.[118–122] This is also exhibited in the hippocampus at subanesthetic doses.[122] However, the translational nature of gamma activity in humans is still unconfirmed. A recent clinical study showed a significant interaction between *baseline* gamma power and mood response. Patients with MDD experienced a better antidepressant response if they exhibited a lower baseline gamma power, whereas patients with a higher baseline gamma power had a worse antidepressant response regardless of the change in gamma power after ketamine administration.[123] This may suggest ketamine's mechanism of action involving restoration of homeostatic balance to induce its antidepressant effect.[123] Patients with higher baseline gamma power may not be receptive to the specific mechanism ketamine utilizes to induce its effects, or experience a further disruption in homeostasis, with no improvements in mood.[123] However, this report had many limitations, as gamma power is highly variable and not an identified biomarker for MDD.[123]

Effects of ketamine also require opioid system activation. Patients with treatment-resistant depression who received ketamine + naltrexone (opioid receptor antagonist) exhibited significantly less improvement in mood than patients who received ketamine + placebo on days 1 and 3 after infusion.[124] Interestingly, there were no differences in dissociation scores, which suggests that the dissociative effects of ketamine are not attributed to the opioid effect. However, with opioid receptor antagonism, effects of ketamine are not enough to produce the rapid antidepressant effect.[124] Ketamine's properties at the opioid receptor represents a public health significance and its potential to be a drug of abuse.[124]

These reports together exemplify the limitations of molecular neuroplasticity being ketamine's sole mechanism of action to produce rapid antidepressant effects. More importantly, they highlight how little we know about ketamine's effect and which mechanism is attributed to induce the rapid antidepressant response.

Conclusion

This chapter summarizes evidence from preclinical and clinical studies to highlight that rapid restoration and alteration of molecular neuroplasticity are likely involved in inducing ketamine's rapid antidepressant effects. Specific molecules of interest include glutamate, AMPAR, mTOR, BDNF/TrkB, VGF, eEF2K, p70S6K, GSK-3, IGF-2, Erk, and microRNAs. These mechanisms seem to diverge from those involved in ketamine's sustained effect, and they do not seem to work independently, as briefly discussed. Further studies are required to explore and clarify which of the discussed events are mainly responsible for ketamine's rapid antidepressant mechanism.

References

1. James SL, Abate D, Abate KH, et al. Global, regional, and national incidence, prevalence, and years lived with disability for 354 diseases and injuries for 195 countries and territories, 1990–2017: a systematic analysis for the Global Burden of Disease Study 2017. *The Lancet*. 2018;392(10159):1789–1858.
2. Fava M, Davidson KG. Definition and epidemiology of treatment-resistant depression. *Psychiatr Clin*. 1996;19:179–200.
3. Phillips JL, Norris S, Talbot J, et al. Single, repeated, and maintenance ketamine infusions for treatment-resistant depression: a randomized controlled trial. *Am J Psychiatr*. 2019;176(5):401–409. https://doi.org/10.1176/appi.ajp.2018.18070834.
4. Murrough JW, Perez AM, Pillemer S, et al. Rapid and longer-term antidepressant effects of repeated ketamine infusions in treatment-resistant major depression. *Biol Psychiatr*. 2013;74(4):250–256. https://doi.org/10.1016/j.biopsych.2012.06.022.
5. aan het Rot M, Zarate CA, Charney DS, Mathew SJ. Ketamine for depression: where do we go from here? *Biol Psychiatr*. 2012;72(7):537–547. https://doi.org/10.1016/j.biopsych.2012.05.003.
6. Murrough JW. Ketamine as a novel antidepressant: from synapse to behavior. *Clin Pharmacol Ther*. 2012;91(2):303–309. https://doi.org/10.1038/clpt.2011.244.
7. Berman RM, Cappiello A, Anand A, et al. Antidepressant effects of ketamine in depressed patients. *Biol Psychiatr*. 2000;47(4):351–354.
8. Zarate CA, Singh JB, Carlson PJ, et al. A randomized trial of an N-methyl-D-aspartate antagonist in treatment-resistant major depression. *Arch Gen Psychiatr*. 2006;63(8):856–864. https://doi.org/10.1001/archpsyc.63.8.856.
9. Short B, Fong J, Galvez V, Shelker W, Loo CK. Side-effects associated with ketamine use in depression: a systematic review. *Lancet Psychiatr*. 2018;5(1):65–78. https://doi.org/10.1016/S2215-0366(17)30272-9.
10. Jansen KL. Non-medical use of ketamine. *Br Med J*. 1993;306(6878):601–602.
11. Price RB, Duman R. Neuroplasticity in cognitive and psychological mechanisms of depression: an integrative model. *Mol Psychiatr*. 2019;1–14. https://doi.org/10.1038/s41380-019-0615-x.
12. Duman RS, Aghajanian GK. Synaptic dysfunction in depression: potential therapeutic targets. *Science*. 2012;338(6103):68–72. https://doi.org/10.1126/science.1222939.

13. Cramer SC, Sur M, Dobkin BH, et al. Harnessing neuroplasticity for clinical applications. *Brain J Neurol*. 2011;134(Pt 6):1591–1609. https://doi.org/10.1093/brain/awr039.
14. Pittenger C, Duman RS. Stress, depression, and neuroplasticity: a convergence of mechanisms. *Neuropsychopharmacology*. 2008;33(1):88–109. https://doi.org/10.1038/sj.npp.1301574.
15. Pascual-Leone A, Amedi A, Fregni F, Merabet LB. The plastic human brain cortex. *Annu Rev Neurosci*. 2005;28(1):377–401. https://doi.org/10.1146/annurev.neuro.27.070203.144216.
16. Koolschijn PCMP, van Haren NEM, Lensvelt-Mulders GJLM, Hulshoff Pol HE, Kahn RS. Brain volume abnormalities in major depressive disorder: a meta-analysis of magnetic resonance imaging studies. *Hum Brain Mapp*. 2009;30(11):3719–3735. https://doi.org/10.1002/hbm.20801.
17. Duman RS, Malberg J, Nakagawa S, D'Sa C. Neuronal plasticity and survival in mood disorders. *Biol Psychiatr*. 2000;48(8):732–739.
18. Duman RS, Aghajanian GK, Sanacora G, Krystal JH. Synaptic plasticity and depression: new insights from stress and rapid-acting antidepressants. *Nat Med*. 2016;22(3): 238–249. https://doi.org/10.1038/nm.4050.
19. Bessa JM, Ferreira D, Melo I, et al. Hippocampal neurogenesis induced by antidepressant drugs: an epiphenomenon in their mood-improving actions. *Mol Psychiatr*. 2009; 14(8):739. https://doi.org/10.1038/mp.2009.75.
20. D'Sa C, Duman RS. Antidepressants and neuroplasticity. *Bipolar Disord*. 2002;4(3): 183–194.
21. Miller OH, Moran JT, Hall BJ. Two cellular hypotheses explaining the initiation of ketamine's antidepressant actions: direct inhibition and disinhibition. *Neuropharmacology*. 2016;100:17–26. https://doi.org/10.1016/j.neuropharm.2015.07.028.
22. Zanos P, Gould TD. Mechanisms of ketamine action as an antidepressant. *Mol Psychiatr*. 2018;23(4):801–811. https://doi.org/10.1038/mp.2017.255.
23. Homayoun H, Moghaddam B. NMDA receptor hypofunction produces opposite effects on prefrontal cortex interneurons and pyramidal neurons. *J Neurosci*. 2007;27(43): 11496–11500. https://doi.org/10.1523/JNEUROSCI.2213-07.2007.
24. Widman AJ, McMahon LL. Disinhibition of CA1 pyramidal cells by low-dose ketamine and other antagonists with rapid antidepressant efficacy. *Proc Natl Acad Sci USA*. 2018; 115(13):E3007–E3016. https://doi.org/10.1073/pnas.1718883115.
25. Abdallah CProc Natl Acad Sci USAG, Jackowski A, Salas R, et al. The nucleus accumbens and ketamine treatment in major depressive disorder. *Neuropsychopharmacology*. 2017;42(8):1739–1746. https://doi.org/10.1038/npp.2017.49.
26. Shungu DC, Milak MS, Kegeles LS, Proper CJ, Mao X, Mann JJ. In vivo human brain 1H MRS monitoring of dynamic amino acid neurotransmitter response to acute administration of ketamine. *Mol Imag Biol*. 2010;12(suppl. 2):S669. https://doi.org/10.1007/s11307-010-0453-3.
27. Milak MS, Proper CJ, Mulhern ST, et al. A pilot in vivo proton magnetic resonance spectroscopy study of amino acid neurotransmitter response to ketamine treatment of major depressive disorder. *Mol Psychiatr*. 2016;21(3):320–327. https://doi.org/10.1038/mp.2015.83.
28. Scheidegger M, Henning A, Fuchs A, et al. Effects of an antidepressant dose of ketamine on prefrontal aspartate, glutamine and gaba levels in healthy subjects: assessing

the post-infusion interval with 1H-MRS. *Biol Psychiatr.* 2013;73(9 suppl. 1): 190S–191S.
29. Chowdhury GMI, Zhang J, Thomas M, et al. Transiently increased glutamate cycling in rat PFC is associated with rapid onset of antidepressant-like effects. *Mol Psychiatr.* 2017;22(1):120–126. https://doi.org/10.1038/mp.2016.34.
30. Wang N, Zhang G-F, Liu X-Y, et al. Downregulation of neuregulin 1-ErbB4 signaling in parvalbumin interneurons in the rat brain may contribute to the antidepressant properties of ketamine. *J Mol Neurosci.* 2014. https://doi.org/10.1007/s12031-014-0277-8.
31. Pham TH, Defaix C, Xu X, et al. Common neurotransmission recruited in (R,S)-Ketamine and (2R,6R)-hydroxynorketamine–induced sustained antidepressant-like effects. *Biol Psychiatr.* 2018;84(1):e3–e6. https://doi.org/10.1016/j.biopsych.2017.10.020.
32. Zhu X, Ye G, Wang Z, Luo J, Hao X. Sub-anesthetic doses of ketamine exert antidepressant-like effects and upregulate the expression of glutamate transporters in the hippocampus of rats. *Neurosci Lett.* 2017;639:132–137. https://doi.org/10.1016/j.neulet.2016.12.070.
33. Boczek T, Lisek M, Ferenc B, Wiktorska M, Ivchevska I, Zylinska L. Region-specific effects of repeated ketamine administration on the presynaptic GABAergic neurochemistry in rat brain. *Neurochem Int.* 2015;91:13–25. https://doi.org/10.1016/j.neuint.2015.10.005.
34. Njau S, Vasavada M, Leaver A, et al. Gabaergic neurotransmission modulates therapeutic response to ketamine infusion in major depression. *Biol Psychiatr.* 2018;83(9 suppl 1):S296–S297.
35. Evans JW, Lally N, An L, et al. 7T 1H-MRS in major depressive disorder: a Ketamine Treatment Study. *Neuropsychopharmacology.* 2018;43(9):1908–1914. https://doi.org/10.1038/s41386-018-0057-1.
36. Li M, Woelfer M, Colic L, et al. Default mode network connectivity change corresponds to ketamine's delayed glutamatergic effects. *Eur. Arch. Psychiatry Clin. Neurosci.*. 2018 https://doi.org/10.1007/s00406-018-0942-y (Li, Woelfer, Colic, Walter) Clinical Affective Neuroimaging Laboratory, Leibniz Institute for Neurobiology, Magdeburg, Germany.
37. Hollmann M, O'Shea-Greenfield A, Rogers SW, Heinemann S. Cloning by functional expression of a member of the glutamate receptor family. *Nature.* 1989;342(6250): 643–648. https://doi.org/10.1038/342643a0.
38. Nakanishi N, Shneider NA, Axel R. A family of glutamate receptor genes: evidence for the formation of heteromultimeric receptors with distinct channel properties. *Neuron.* 1990;5(5):569–581. https://doi.org/10.1016/0896-6273(90)90212-X.
39. Raymond LA, Blackstone CD, Huganir RL. Phosphorylation of amino acid neurotransmitter receptors in synaptic plasticity. *Trends Neurosci.* 1993;16(4):147–153. https://doi.org/10.1016/0166-2236(93)90123-4.
40. Maeng S, Zarate Jr CA, Du J, et al. Cellular mechanisms underlying the antidepressant effects of ketamine: role of alpha-amino-3-hydroxy-5-methylisoxazole-4-propionic acid receptors. *Biol Psychiatr.* 2008;63(4):349–352. https://doi.org/10.1016/j.biopsych.2007.05.028.
41. Zhou W, Wang N, Yang C, Li X-M, Zhou Z-Q, Yang J-J. Ketamine-induced antidepressant effects are associated with AMPA receptors-mediated upregulation of mTOR and BDNF in rat hippocampus and prefrontal cortex. *Eur Psychiatr.* 2014;29(7):419–423. https://doi.org/10.1016/j.eurpsy.2013.10.005.

42. Manji H, Du J, Chen G, Zarate CA. Regulating AMPA/NMDA mediated synaptic plasticity in critical circuits for the development of novel, improved rapidly acting therapeutics. *Int J Neuropsychopharmacol*. 2010;13(suppl. 1):12. https://doi.org/10.1017/S1461145710000635.
43. Akinfiresoye L, Tizabi Y. Antidepressant effects of AMPA and ketamine combination: role of hippocampal BDNF, synapsin, and mTOR. *Psychopharmacology*. 2013;230(2): 291–298. https://doi.org/10.1007/s00213-013-3153-2.
44. Beurel E, Grieco SF, Amadei C, Downey K, Jope RS. Ketamine-induced inhibition of glycogen synthase kinase-3 contributes to the augmentation of α-amino- 3- hydroxy-5-methylisoxazole-4-propionic acid (AMPA) receptor signaling. *Bipolar Disord*. 2016; 18(6):473–480. https://doi.org/10.1111/bdi.12436.
45. Duman R. Rapid antidepressant actions of ketamine require stimulation of mammalian target of rapaymicin (mTOR) signaling and synaptic protein synthesis. *Neuropsychopharmacology*. 2010;35(suppl. 1):S18–S19. https://doi.org/10.1038/npp.2010.215.
46. Lepack AE, Bang E, Lee B, Dwyer JM, Duman RS. Fast-acting antidepressants rapidly stimulate ERK signaling and BDNF release in primary neuronal cultures. *Neuropharmacology*. 2016;111:242–252. https://doi.org/10.1016/j.neuropharm.2016.09.011.
47. Girgenti MJ, Ghosal S, LoPresto D, Taylor JR, Duman RS. Ketamine accelerates fear extinction via mTORC1 signaling. *Neurobiol Dis*. 2017:1–8. https://doi.org/10.1016/j.nbd.2016.12.026.
48. Llamosas N, Perez-Caballero L, Berrocoso E, Bruzos-Cidon C, Ugedo L, Torrecilla M. Ketamine promotes rapid and transient activation of AMPA receptor-mediated synaptic transmission in the dorsal raphe nucleus. *Prog Neuro-Psychopharmacol Biol Psychiatry*. 2019;88:243–252.
49. Tizabi Y, Bhatti BH, Manaye KF, Das JR, Akinfiresoye L. Antidepressant-like effects of low ketamine dose is associated with increased hippocampal AMPA/NMDA receptor density ratio in female Wistar-Kyoto rats. *Neuroscience*. 2012;213:72–80. https://doi.org/10.1016/j.neuroscience.2012.03.052.
50. Ho M-F, Correia C, Ingle JN, et al. Ketamine and ketamine metabolites as novel estrogen receptor ligands: induction of cytochrome P450 and AMPA glutamate receptor gene expression. *Biochem Pharmacol*. 2018;152:279–292. https://doi.org/10.1016/j.bcp.2018.03.032.
51. Zhang K, Xu T, Yuan Z, et al. Essential roles of AMPA receptor GluA1 phosphorylation and presynaptic HCN channels in fast-Acting antidepressant responses of ketamine. *Sci Signal*. 2016;9(458):ra123. https://doi.org/10.1126/scisignal.aai7884.
52. Hay N, Sonenberg N. Upstream and downstream of mTOR. *Genes Dev*. 2004;18(16): 1926–1945. https://doi.org/10.1101/gad.1212704.
53. Laplante M, Sabatini DM. MTOR signaling in growth control and disease. *Cell*. 2012; 149(2):274–293. https://doi.org/10.1016/j.cell.2012.03.017.
54. Li N, Lee B, Liu R-J, et al. MTOR-dependent synapse formation underlies the rapid antidepressant effects of NMDA antagonists. *Science*. 2010;329(5994):959–964. https://doi.org/10.1126/science.1190287.
55. Li N, Liu R-J, Dwyer JM, et al. Glutamate N-methyl-D-aspartate receptor antagonists rapidly reverse behavioral and synaptic deficits caused by chronic stress exposure. *Biol Psychiatr*. 2011;69(8):754–761. https://doi.org/10.1016/j.biopsych.2010.12.015.
56. Gerhard DM, Wohleb ES, Duman RS. A Role for NR2B-Containing NMDA receptors on gabaergic interneurons in the rapid antidepressant effects of ketamine. *Biol Psychiatr*. 2016;79(9 suppl. 1):63S. https://doi.org/10.1016/j.biopsych.2016.03.1748.

57. Miller OH, Yang L, Wang C-C, et al. GluN2B-containing NMDA receptors regulate depression-like behavior and are critical for the rapid antidepressant actions of ketamine. *Elife*. 2014;3:e03581. https://doi.org/10.7554/eLife.03581.
58. Harraz MM, Tyagi R, Cortés P, Snyder SH. Antidepressant action of ketamine via mTOR is mediated by inhibition of nitrergic Rheb degradation. *Mol Psychiatr.* 2016; 21(3):313–319. https://doi.org/10.1038/mp.2015.211.
59. Castrén E. Neurotrophic effects of antidepressant drugs. *Curr Opin Pharmacol*. 2004; 4(1):58–64. https://doi.org/10.1016/j.coph.2003.10.004.
60. Chen B, Dowlatshahi D, MacQueen GM, Wang J-F, Young LT. Increased hippocampal bdnf immunoreactivity in subjects treated with antidepressant medication. *Biol Psychiatr.* 2001;50(4):260–265. https://doi.org/10.1016/S0006-3223(01)01083-6.
61. Cavalleri L, Merlo Pich E, Millan MJ, et al. Ketamine enhances structural plasticity in mouse mesencephalic and human iPSC-derived dopaminergic neurons via AMPAR-driven BDNF and mTOR signaling. *Mol Psychiatr.* 2018;23(4):812–823. https://doi.org/10.1038/mp.2017.241.
62. Cunha C, Brambilla R, Thomas KL. A simple role for BDNF in learning and memory? *Front Mol Neurosci*. 2010;3. https://doi.org/10.3389/neuro.02.001.2010.
63. Xu SX, Zhou ZQ, Li XM, Ji MH, Zhang GF, Yang JJ. The activation of adenosine monophosphate-activated protein kinase in rat hippocampus contributes to the rapid antidepressant effect of ketamine. *Behav Brain Res*. 2013;253:305–309. https://doi.org/10.1016/j.bbr.2013.07.032.
64. Zhang M, Radford KD, Driscoll M, Purnomo S, Kim J, Choi KH. Effects of subanesthetic intravenous ketamine infusion on neuroplasticity-related proteins in the prefrontal cortex, amygdala, and hippocampus of Sprague-Dawley rats. *IBRO Rep*. 2019;6:87–94. https://doi.org/10.1016/j.ibror.2019.01.006.
65. Liu R-J, Fuchikami M, Dwyer JM, Lepack AE, Duman RS, Aghajanian GK. GSK-3 inhibition potentiates the synaptogenic and antidepressant-like effects of subthreshold doses of ketamine. *Neuropsychopharmacology.* 2013;38(11):2268–2277. https://doi.org/10.1038/npp.2013.128.
66. Yang C, Hu Y-M, Zhou Z-Q, Zhang G-F, Yang J-J. Acute administration of ketamine in rats increases hippocampal BDNF and mTOR levels during forced swimming test. *Ups J Med Sci*. 2013;118(1):3–8. https://doi.org/10.3109/03009734.2012.724118.
67. Silva Pereira V, Elfving B, Joca SRL, Wegener G. Ketamine and aminoguanidine differentially affect Bdnf and Mtor gene expression in the prefrontal cortex of adult male rats. *Eur J Pharmacol*. 2017;815:304–311. https://doi.org/10.1016/j.ejphar.2017.09.029.
68. Lepack AE, Fuchikami M, Dwyer JM, Banasr M, Duman RS. BDNF release is required for the behavioral actions of ketamine. *Int J Neuropsychopharmacol*. 2014;18(1). https://doi.org/10.1093/ijnp/pyu033.
69. Autry AE, Adachi M, Nosyreva E, et al. NMDA receptor blockade at rest triggers rapid behavioural antidepressant responses. *Nature*. 2011;475(7354):91–96. https://doi.org/10.1038/nature10130.
70. Liu W-X, Wang J, Xie Z-M, et al. Regulation of glutamate transporter 1 via BDNF-TrkB signaling plays a role in the anti-apoptotic and antidepressant effects of ketamine in chronic unpredictable stress model of depression. *Psychopharmacology.* 2016; 233(3):405–415. https://doi.org/10.1007/s00213-015-4128-2.
71. Ma Z, Zang T, Birnbaum SG, et al. TrkB dependent adult hippocampal progenitor differentiation mediates sustained ketamine antidepressant response. *Nat Commun*. 2017; 8(1):1668. https://doi.org/10.1038/s41467-017-01709-8.

72. Zhang F, Luo J, Zhu X. Ketamine ameliorates depressive-like behaviors by tPA-mediated conversion of proBDNF to mBDNF in the hippocampus of stressed rats. *Psychiatr Res*. 2018;269:646–651. https://doi.org/10.1016/j.psychres.2018.08.075.
73. Fraga DB, Reus GZ, Abelaira HM, et al. Ketamine alters behavior and decreases BDNF levels in the rat brain as a function of time after drug administration. *Rev Bras Psiquiatr.* 2013;35(3):262–266. https://doi.org/10.1590/1516-4446-2012-0858.
74. Garcia LSB, Comim CM, Valvassori SS, et al. Acute administration of ketamine induces antidepressant-like effects in the forced swimming test and increases BDNF levels in the rat hippocampus. *Prog Neuro-Psychopharmacol Biol Psychiatry*. 2008; 32(1):140–144. https://doi.org/10.1016/j.pnpbp.2007.07.027.
75. Garcia LS, Comim CM, Valvassori SS, et al. Chronic administration of ketamine elicits antidepressant-like effects in rats without affecting hippocampal brain-derived neurotrophic factor protein levels. *Basic Clin Pharmacol Toxicol*. 2008;103(6):502–506. https://doi.org/10.1111/j.1742-7843.2008.00210.x.
76. Garcia LS, Quevedo J, Comim CM, et al. Ketamine treatment reverses behavioral and physiological alterations induced by chronic mild stress in rats. *Bipolar Disord*. 2009; 11(S1):71. https://doi.org/10.1016/j.pnpbp.2009.01.004.
77. Le Nedelec M, Glue P, Winter H, Goulton C, Broughton L, Medlicott N. Acute low-dose ketamine produces a rapid and robust increase in plasma BDNF without altering brain BDNF concentrations. *Drug Deliv Transl Res*. 2018;8(3):780–786. https://doi.org/10.1007/s13346-017-0476-2.
78. Allen AP, Naughton M, Dowling J, et al. Serum BDNF as a peripheral biomarker of treatment-resistant depression and the rapid antidepressant response: a comparison of ketamine and ECT. *J Affect Disord*. 2015;186:306–311. https://doi.org/10.1016/j.jad.2015.06.033.
79. Duncan Jr WC, Sarasso S, Ferrarelli F, et al. Concomitant BDNF and sleep slow wave changes indicate ketamine-induced plasticity in major depressive disorder. *Int J Neuropsychopharmacol*. 2013;16(2):301–311. https://doi.org/10.1017/S1461145712000545.
80. Haile CN, Murrough JW, Iosifescu DV, et al. Plasma brain derived neurotrophic factor (BDNF) and response to ketamine in treatment-resistant depression. *Int J Neuropsychopharmacol*. 2014;17(2):331–336. https://doi.org/10.1017/S1461145713001119.
81. Machado-Vieira R, Yuan P, Brutsche N, et al. Brain-derived neurotrophic factor and initial antidepressant response to an N-methyl-D-aspartate antagonist. *J Clin Psychiatr.* 2009;70(12):1662–1666. https://doi.org/10.4088/JCP.08m04659.
82. Hunsberger JG, Newton SS, Bennett AH, et al. Antidepressant actions of the exercise-regulated gene VGF. *Nat Med*. 2007;13(12):1476–1482. https://doi.org/10.1038/nm1669.
83. Jiang C, Lin W-J, Sadahiro M, et al. VGF function in depression and antidepressant efficacy. *Mol Psychiatr.* 2018;23(7):1632–1642. https://doi.org/10.1038/mp.2017.233.
84. Jiang C, Lin W-J, Labonte B, et al. VGF and its C-terminal peptide TLQP-62 in ventromedial prefrontal cortex regulate depression-related behaviors and the response to ketamine. *Neuropsychopharmacology*. 2018. https://doi.org/10.1038/s41386-018-0277-4 (Jiang, Lin, Labonte, Nestler, Russo, Salton) Department of Neuroscience, Icahn School of Medicine at Mount Sinai, New York, NY 10029, United States.
85. Monteggia LM, Gideons E, Kavalali ET. The role of eukaryotic elongation factor 2 kinase in rapid antidepressant action of ketamine. *Biol Psychiatr.* 2013;73(12): 1199–1203. https://doi.org/10.1016/j.biopsych.2012.09.006.

86. Sutton MA, Taylor AM, Ito HT, Pham A, Schuman EM. Postsynaptic decoding of neural activity: EEF2 as a biochemical sensor coupling miniature synaptic transmission to local protein synthesis. *Neuron*. 2007;55(4):648–661. https://doi.org/10.1016/j.neuron.2007.07.030.
87. Abelaira HM, Reus GZ, Ignacio ZM, et al. Effects of ketamine administration on mTOR and reticulum stress signaling pathways in the brain after the infusion of rapamycin into prefrontal cortex. *J Psychiatr Res*. 2017;87:81–87. https://doi.org/10.1016/j.jpsychires.2016.12.002.
88. Chiamulera C, Di Chio M, Cavalleri L, Venniro M, Padovani L, Collo G. Ketamine effects on mammalian target of rapamycin signaling in the mouse limbic system depend on functional dopamine D3 receptors. *NeuroRep*. 2018;29(8):615–620. https://doi.org/10.1097/WNR.0000000000001008.
89. Frame S, Cohen P. GSK3 takes centre stage more than 20 years after its discovery. *Biochem J*. 2001;359(Pt 1):1–16.
90. Beurel E, Song L, Jope RS. Inhibition of glycogen synthase kinase-3 is necessary for the rapid antidepressant effect of ketamine in mice. *Mol Psychiatr*. 2011;16(11):1068–1070. https://doi.org/10.1038/mp.2011.47.
91. Grieco SF, Cheng Y, Eldar-Finkelman H, Jope RS, Beurel E. Up-regulation of insulin-like growth factor 2 by ketamine requires glycogen synthase kinase-3 inhibition. *Prog Neuro-Psychopharmacol Biol Psychiatry*. 2017;72:49–54. https://doi.org/10.1016/j.pnpbp.2016.08.008.
92. Zhou W, Dong L, Wang N, et al. Akt mediates GSK-3β phosphorylation in the rat prefrontal cortex during the process of ketamine exerting rapid antidepressant actions. *Neuroimmunomodulation*. 2014b;21(4):183–188. https://doi.org/10.1159/000356517.
93. Chen X, Nelson CD, Li X, et al. PSD-95 is required to sustain the molecular organization of the postsynaptic density. *J Neurosci*. 2011;31(17):6329–6338. https://doi.org/10.1523/JNEUROSCI.5968-10.2011.
94. Zhang W-J, Wang H-H, Lv Y-D, Liu C-C, Sun W-Y, Tian L-J. Downregulation of egr-1 expression level via GluN2B underlies the antidepressant effects of ketamine in a chronic unpredictable stress animal model of depression. *Neuroscience*. 2018;372:38–45. https://doi.org/10.1016/j.neuroscience.2017.12.045.
95. Carrier N, Kabbaj M. Sex differences in the antidepressant-like effects of ketamine. *Neuropharmacology*. 2013;70:27–34. https://doi.org/10.1016/j.neuropharm.2012.12.009.
96. Thelen C, Flaherty E, Saurine J, Sens J, Mohamed S, Pitychoutis PM. Sex differences in the temporal neuromolecular and synaptogenic effects of the rapid-acting antidepressant drug ketamine in the mouse brain. *Neuroscience*. 2018;398:182–192. https://doi.org/10.1016/j.neuroscience.2018.11.053.
97. Colic L, McDonnell C, Li M, et al. Neuronal glutamatergic changes and peripheral markers of cytoskeleton dynamics change synchronically 24h after sub-anaesthetic dose of ketamine in healthy subjects. *Behav Brain Res*. 2019;359:312–319. https://doi.org/10.1016/j.bbr.2018.10.021.
98. Walker AJ, Foley BM, Hu C, Sutor SL, Frye MA, Tye SJ. Preclinical infralimbic mTOR signaling correlates with treatment response to ketamine. *Biol Psychiatr*. 2014;75(9 suppl. 1):312S. https://doi.org/10.1016/j.biopsych.2014.03.016.
99. Abdallah CG, Averill LA, Gueorguieva R, et al. Rapamycin, an immunosuppressant and mTORC1 inhibitor, triples the antidepressant response rate of ketamine at 2 Weeks

following treatment: a double-blind, placebo-controlled, cross-over, randomized clinical trial. *bioRxiv*. 2018:500959. https://doi.org/10.1101/500959.

100. Wray NH, Schappi JM, Singh H, Senese NB, Rasenick MM. NMDAR-independent, cAMP-dependent antidepressant actions of ketamine. *Mol Psychiatr*. 2018;Vols. 1–11. https://doi.org/10.1038/s41380-018-0083-8 (Wray, Rasenick) Graduate Program in Neuroscience, University of Illinois at Chicago, Chicago, IL, United States.
101. Reus GZ, Abaleira HM, Titus SE, et al. Effects of ketamine administration on the phosphorylation levels of CREB and TrKB and on oxidative damage after infusion of MEK inhibitor. *Pharmacol Rep*. 2016;68(1):177–184. https://doi.org/10.1016/j.pharep.2015.08.010.
102. Deyama S, Bang E, Wohleb ES, et al. Role of neuronal VEGF signaling in the prefrontal cortex in the rapid antidepressant effects of ketamine. *Am J Psychiatr*. 2019. https://doi.org/10.1176/appi.ajp.2018.17121368.
103. du Jardin KG, Müller HK, Sanchez C, Wegener G, Elfving B. A single dose of vortioxetine, but not ketamine or fluoxetine, increases plasticity-related gene expression in the rat frontal cortex. *Eur J Pharmacol*. 2016;786:29–35. https://doi.org/10.1016/j.ejphar.2016.05.029.
104. Wan Y-Q, Feng J-G, Li M, et al. Prefrontal cortex miR-29b-3p plays a key role in the antidepressant-like effect of ketamine in rats. *Exp Mol Med*. 2018;50(10):140. https://doi.org/10.1038/s12276-018-0164-4.
105. Yang X, Yang Q, Wang X, et al. MicroRNA expression profile and functional analysis reveal that miR-206 is a critical novel gene for the expression of BDNF induced by ketamine. *NeuroMolecular Med*. 2014;16(3):594–605. https://doi.org/10.1007/s12017-014-8312-z.
106. Yang C, Zhang J-C, Ren Q, et al. R-Ketamine: a rapid onset and sustained antidepressant without psychotomimetic side effects. *Trans Psychiatr*. 2015;5:e632. https://doi.org/10.1038/tp.2015.136.
107. Ng LHL, Huang Y, Han L, Chang RC-C, Chan YS, Lai CSW. Ketamine and selective activation of parvalbumin interneurons inhibit stress-induced dendritic spine elimination. *Transl Psychiatry*. 2018;8(1):272. https://doi.org/10.1038/s41398-018-0321-5.
108. Phoumthipphavong V, Barthas F, Hassett S, Kwan AC. Longitudinal effects of ketamine on dendritic architecture in vivo in the mouse medial frontal cortex. *ENeuro*. 2016;3(2). https://doi.org/10.1523/ENEURO.0133-15.2016.
109. Pryazhnikov E, Mugantseva E, Casarotto P, et al. Longitudinal two-photon imaging in somatosensory cortex of behaving mice reveals dendritic spine formation enhancement by subchronic administration of low-dose ketamine. *Sci Rep*. 2018;8(1):6464. https://doi.org/10.1038/s41598-018-24933-8.
110. Moda-Sava RN, Murdock MH, Parekh PK, et al. Sustained rescue of prefrontal circuit dysfunction by antidepressant-induced spine formation. *Science*. 2019;(6436):364. https://doi.org/10.1126/science.aat8078.
111. Knott GW, Holtmaat A, Wilbrecht L, Welker E, Svoboda K. Spine growth precedes synapse formation in the adult neocortex in vivo. *Nat Neurosci*. 2006;9(9):1117–1124. https://doi.org/10.1038/nn1747.
112. Bonanno G, Giambelli R, Raiteri L, et al. Chronic antidepressants reduce depolarization-evoked glutamate release and protein interactions favoring formation of SNARE complex in Hippocampus. *J Neurosci*. 2005;25(13):3270–3279. https://doi.org/10.1523/JNEUROSCI.5033-04.2005.

113. Barbiero VS, Giambelli R, Musazzi L, et al. Chronic antidepressants induce redistribution and differential activation of αCaM kinase II between presynaptic compartments. *Neuropsychopharmacology*. 2007;32(12):2511–2519. https://doi.org/10.1038/sj.npp.1301378.
114. Nosyreva E, Szabla K, Autry AE, Ryazanov AG, Monteggia LM, Kavalali ET. Acute suppression of spontaneous neurotransmission drives synaptic potentiation. *J Neurosci*. 2013;33(16):6990–7002. https://doi.org/10.1523/JNEUROSCI.4998-12.2013.
115. Abdallah CG, Averill LA, Collins KA, et al. Ketamine treatment and global brain connectivity in major depression. *Neuropsychopharmacology*. 2017;42(6):1210–1219. https://doi.org/10.1038/npp.2016.186.
116. Albott C, Shiroma P, Thuras P, Pardo J, Lim K. The antidepressant effect of repeat dose intravenous ketamine is delayed by concurrent benzodiazepine use. *J Clin Psychiatr*. 2017;78(3):e308–e309. https://doi.org/10.4088/JCP.16l11277.
117. Murrough J, Collins K, Fields J, et al. Regulation of neural responses to emotion by ketamine in individuals with treatment-resistant major depression. *Neuropsychopharmacology*. 2014;39(suppl. 1):S554. https://doi.org/10.1038/npp.2014.282.
118. Amat-Foraster M, Jensen AA, Plath N, Herrik KF, Celada P, Artigas F. Temporally dissociable effects of ketamine on neuronal discharge and gamma oscillations in rat thalamo-cortical networks. *Neuropharmacology*. 2018;137:13–23. https://doi.org/10.1016/j.neuropharm.2018.04.022.
119. Carlén M, Meletis K, Siegle JH, et al. A critical role for NMDA receptors in parvalbumin interneurons for gamma rhythm induction and behavior. *Mol Psychiatr*. 2012;17(5):537–548. https://doi.org/10.1038/mp.2011.31.
120. Lazarewicz MT, Ehrlichman RS, Maxwell CR, Gandal MJ, Finkel LH, Siegel SJ. Ketamine modulates theta and gamma oscillations. *J Cognit Neurosci*. 2010;22(7):1452–1464. https://doi.org/10.1162/jocn.2009.21305.
121. Huang L, Yang X-J, Huang Y, Sun EY, Sun M. Ketamine protects gamma oscillations by inhibiting hippocampal LTD. *PLoS One*. 2016;11(7):e0159192. https://doi.org/10.1371/journal.pone.0159192.
122. Nagy D, Stoiljkovic M, Menniti FS, Hajós M. Differential effects of an NR2B NAM and ketamine on synaptic potentiation and gamma synchrony: relevance to rapid-onset antidepressant efficacy. *Neuropsychopharmacology*. 2016;41(6):1486–1494. https://doi.org/10.1038/npp.2015.298.
123. Nugent AC, Ballard ED, Gould TD, et al. Ketamine has distinct electrophysiological and behavioral effects in depressed and healthy subjects. *Mol Psychiatr*. 2019;24(7):1040–1052. https://doi.org/10.1038/s41380-018-0028-2\.
124. Williams NR, Heifets BD, Blasey C, et al. Opioid receptor antagonism attenuates antidepressant effects of ketamine. *Am J Psychiatr*. 2018;175(12):1205–1215. https://doi.org/10.1176/appi.ajp.2018.18020138.

CHAPTER

Treatment resistant depression

3

Sophie R. Vaccarino, HBSc [1,2], Sidney H. Kennedy, MD [3]

[1]*Research Assistant, Centre for Depression and Suicide Studies, Unity Health Toronto, Toronto, ON, Canada;* [2]*MSc Candidate, Institute of Medical Science, University of Toronto, Toronto, ON, Canada;* [3]*Dr. Psychiatry, University Health Network, Toronto, ON, Canada*

Treatment-resistant depression (TRD) is defined as a failure to achieve an *adequate*, or clinically meaningful, response to an *adequate* trial of antidepressant therapy. Treatment resistance is common among people diagnosed with major depressive disorder (MDD). In fact, it is estimated that approximately 50% of individuals with MDD do not respond to their first antidepressant medication.[1] The likelihood of responding to any given treatment decreases as the number of previously failed trials increases.[2] Furthermore, TRD is associated with significant impairments in psychosocial functioning.[3] Considering the large proportion of individuals with MDD who experience treatment resistance, and the substantial burden TRD places on patients, their families, and the society, an understanding of treatment resistance and potential treatments is critical for clinicians and researchers in psychiatry.

In this chapter, we will review the main approaches to define treatment resistance and the accompanying factors to consider when diagnosing. We will then discuss the different staging approaches that have been proposed to classify levels of treatment resistance. Prevalence of TRD, specifically at different stages of resistance, and risk factors associated with the development of TRD will also be addressed. Finally, we will consider the three major prongs of treatment for TRD: pharmacotherapy, psychotherapy, and neuromodulation. In practice, these treatment strategies are often used in combination; however, we will discuss each individually. Pharmacotherapy will be discussed in terms of monotherapy (i.e., switching medications) and polytherapy (i.e., combining medications). We will also provide a thorough review of new treatment options that are under research as potential alternative strategies for TRD, specifically ketamine and psychedelics.

Defining treatment-resistant depression

Achieving a widely accepted, operational definition for "treatment-resistant depression" has been a challenge for researchers and clinicians for many years. This has created a barrier to compare treatment studies and manage TRD. In its most basic sense, TRD is defined as the failure of an individual with MDD to achieve an

Ketamine for Treatment-Resistant Depression. https://doi.org/10.1016/B978-0-12-821033-8.00003-4

adequate response to an *adequate* trial of antidepressant therapy during a major depressive episode (MDE). What constitutes both an adequate response and an adequate trial is less clear. In fact, neither the *Diagnostic and Statistical Manual of Mental Disorders* (Fifth Edition) (*DSM-5*)[4] nor the *International Classification of Diseases, Eleventh Revision* (*ICD-11*)[5] include TRD as a diagnosis. In a systematic review, 155 different definitions for TRD were identified.[6] The two most frequently applied definitions are based on the number of medication/treatment methods failed or the use of a staging model.[6]

"Adequate response" is typically defined as a 50% or greater reduction in depressive symptoms,[7] based on repeated measures using a depression rating scale such as the Hamilton Depression Rating Scale (HAM-D)[8] or Montgomery-Asberg Depression Rating Scale (MADRS).[9] TRD can be classified as having never achieved an adequate response to an adequate antidepressant trial, or as having relapsed after achieving an adequate response.[6,7]

There is no consensus on what constitutes an "adequate trial": specifically, which different classes of antidepressant agents must be tried? How long is an adequate trial? What is an adequate dose? How is "failure" defined?[6,10,11] However, there are some agreed upon aspects. First, the therapy must be one that has proven efficacy. Furthermore, it must be taken at a stable dose for a period deemed adequate to achieve a therapeutic effect; this period ranges from 2 to 12 weeks across the literature.[11–15] There is disparity in the literature about the number of different types of antidepressants that must be tried before a person is classified as treatment resistant. In clinical research, inclusion criteria typically require participants to have failed two or more different antidepressant therapies, sometimes stipulating different classes of antidepressants.[16] The majority of TRD definitions only consider failure to monotherapy and do not account for psychotherapy, adjunctive pharmacotherapy, or neuromodulation techniques.[6,17,18]

There is no clear consensus on how to define TRD and there is also a discrepancy in the nomenclature that should be used. In one systematic review, a few among the 11 different terms used to describe TRD were "treatment resistant," "medication resistant," and "antidepressant refractory".[17] Although most researchers use the terms "resistant" and "refractory" interchangeably, others differentiate "treatment resistant" as a lack of response to any number of antidepressant trials from "treatment refractory" as no response or worsening of symptoms after a "maximum" number of treatment trials.[17,19]

Staging models

In an attempt to better define, diagnose, and treat TRD, at least five staging models have been proposed: Antidepressant Treatment History Form (ATHF), Thase-Rush Staging Model (TRSM), European Staging Model (ESM), Massachusetts General Hospital Staging (MGH-S) Model, and Maudsley Staging Model (MSM). These staging models are outlined in detail in Table 3.1.

Table 3.1 The five staging models for treatment-resistant depression.

Antidepressant Treatment History Form (ATHF)[13,20]	
0	No psychotropic medication or ECT
1	<4 weeks AD or ≥4 weeks at <50% of an adequate dose 1-3 ECT treatments (unilateral or bilateral) *Psychotic depression:* AD with no AP AP only, <3 weeks or <400 mg/d CPZ Level 1 AD + AP <3 weeks or <400 mg/d CPZ
2	≥4 weeks AD at ≥50% but <100% of an adequate dose 4-6 ECT treatments (unilateral or bilateral) *Psychotic depression:* Level 2-5 AD + AP <3 weeks or <400 mg/d CPZ Level 1-2 AD + AP ≥3 weeks and ≥400 mg/d CPZ AP only ≥3 weeks and ≥400 mg/d CPZ
3	≥4 weeks AD at minimum adequate dose 7-9 unilateral ECT treatments *Psychotic depression:* Level 3 AD + AP ≥3 weeks and ≥400 mg/d CPZ
4	≥4 weeks AD above minimum adequate dose Level 3 AD augmented with lithium for ≥2 weeks 10-12 unilateral or 7-9 bilateral ECT treatments *Psychotic depression:* Level 4 AD + AP ≥3 weeks and ≥400 mg/d CPZ
5	Level 4 AD augmented with lithium for ≥2 weeks ≥13 unilateral or ≥10 bilateral ECT treatments *Psychotic depression:* Level 5 AD + AP ≥3 weeks and ≥400 mg/d CPZ
Thase-Rush Staging Model (TRSM)[21]	
Stage I	Failure of at least 1 adequate trial of 1 major class of AD
Stage II	Failure of at least 2 adequate trials of at least 2 distinctly different classes of ADs
Stage III	Stage II resistance plus failure of an adequate trial of a TCA
Stage IV	Stage III resistance plus failure of an adequate trial of an MAOI
Stage V	Stage IV resistance plus a course of bilateral ECT

Continued

Table 3.1 The five staging models for treatment-resistant depression.—*cont'd*

European Staging Model (ESM)[14]	
Non Responder	Nonresponse to an adequate trial of a TCA, SSRI, MAOI, SNRI, ECT, or other antidepressant *Duration of trial:* 6-8 weeks
Treatment Resistant Depression	Resistance to ≥2 adequate AD trial *Duration of trial(s):* TRD 1: 12-16 weeks TRD 2: 18-24 weeks TRD 3: 24-32 weeks TRD 4: 30-40 weeks TRD 4: 36 weeks - 1 year
Chronic Resistant Depression	Resistance to several AD trials, including an augmentation strategy *Duration of trial(s):* ≥1 year

Massachusetts General Hospital Staging Model (MGH-s)[11]	
Trial of ≥6 weeks of an adequate dose of a marketed AD	1 point for each trial
Optimization of dose, optimization of duration, and augmentation/combination strategy of each trial	0.5 points per optimization/strategy
ECT	3 points

Maudsley Staging Model (MSM)[15]		
Parameter/Dimension	Parameter specification	Score
Duration	Acute (≤12 months)	1
	Sub-acute (13-24 months)	2
	Chronic (>24 months)	3
Symptom severity (at baseline)	Subsyndromal	1
	Syndromal:	
	Mild	2
	Moderate	3
	Severe without psychosis	4
	Severe with psychosis	5
Treatment failures		
Antidepressants	Level 1: 1-2 medications	1
	Level 2: 3-4 medications	2
	Level 3: 5-6 medications	3
	Level 4: 7-10 medications	4
	Level 5: >10 medications	5
Augmentation	Not used	0
	Used	1
ECT	Not used	0
	Used	1
TOTAL		Out of 15

AD, *antidepressant medication;* AP, *antipsychotic medication;* CPZ, *chlorpromazine equivalent;* ECT, *electroconvulsive therapy;* MAOI, *monoamine oxidase inhibitor;* mg/d, *milligrams per day;* SNRI, *serotonin-norepinephrine reuptake inhibitor;* SSRI, *selective serotonin reuptake inhibitor;* TCA, *tricyclic antidepressant;* TRD, *treatment resistant depression.*

Comparing staging models

The ATHF was the first staging model developed (1990, updated 1999) with the goal of describing the adequacy of treatment given before a trial of electroconvulsive therapy (ECT).[13,20] The ATHF gives clear guidelines of the dosage and duration required at each stage of treatment resistance.[13,20] However, the ATHF only accounts for treatment being taken at the time of the interview and does not address polytherapy or augmentation strategies (with the exception of lithium).[13,20]

In contrast, the TRSM (1997) classifies patients based on treatment trials completed and failed.[21] Although the TRSM has the advantage of being very easy to use, it has been heavily criticized for a variety of reasons.[18,21] First, the TRSM provides no explicit definition for "adequate dose" or "adequate trial." This model also assumes a hierarchy of antidepressant classes, implying monoamine oxidase inhibitors (MAOIs) and tricyclic antidepressants (TCAs) are more effective than selective serotonin reuptake inhibitors (SSRIs) and serotonin-norepinephrine reuptake inhibitors (SNRIs) for severe TRD and that it is more difficult to treat a patient who has failed two trials with different antidepressant classes than with two or more antidepressants of the same class.[18,21] Finally, the TRSM does not account for adjunctive strategies,[21] which would typically be employed in clinical practice before TCAs and/or MAOIs are prescribed.

To address the limitations of the TRSM, the MGH-S Model (2003) was developed.[11] The MGH-S version gives a clear definition of adequate trial duration (at least 6 weeks), does not assume a hierarchy of antidepressants (all classes are rated equally), and accounts for optimization and adjunctive strategies employed.[11]

The ESM (1999) and MSM (2008) take into account not only the antidepressant trials failed but also individual patient features. The ESM classifies patients as nonresponder, treatment-resistant, or chronic-resistant based on the total duration of their treatment trial(s), while the MSM classifies patients based on the duration of their MDE and symptom severity.

Prevalence

Approximately 40%–50% of individuals with MDD do not respond to their first prescribed antidepressant treatment,[1,2,22] and 20%–30% do not respond to at least two treatment trials.[2,23] The highest rates of TRD are reported in academic/tertiary care settings, followed by inpatient psychiatry units, outpatient units, and primary care settings.[1] The likelihood of responding to an antidepressant decreases with each additional antidepressant trial. Although an individual's likelihood of responding to his/her first antidepressant is 50%–60%, 70% of patients who fail one trial will fail the second and 80%–85% of patients will fail their third or fourth trial when they fail to respond to previous trials.[1,2,22] Furthermore, individuals requiring multiple stages of treatment are more likely to relapse: the relapse rate is 40% after 4 months if a patient responded to his/her first antidepressant trial, but it increases to 55% if two treatments were tried.[2] Relapse rates among individuals requiring three to four trials are approximately 65%–70%.[2]

Diagnosis and differential diagnosis

When a patient presents with resistance to an antidepressant, the recommended first treatment approach is optimization[12]: usually increasing (but occasionally decreasing) the dosage of the current antidepressant to a safe and tolerated level for 6–12 weeks.[11,24] This may involve increasing to "supratherapeutic doses," especially if the patient has required high doses of other medications in the past or has demonstrated good tolerability and partial response to the drug in question.[25] If drug optimization is unsuccessful, clinicians should attempt to rule out other potential causes for the persistence of an MDE despite an adequate therapeutic trial. Before confirming a diagnosis of TRD and potentially switching or adding medication, drug adherence and pharmacokinetic factors, as well as medical and psychiatric comorbidities, should be assessed.

There are many factors that may contribute to nonadherence and an open dialogue between the clinician and patient is central to ensure medication adherence. Reasons for medication nonadherence include the patient's negative perceptions of psychiatric medication, financial constraints, cognitive deficits (i.e., "forgetting"), or medication side effects.[26]

Individual genetic variations that affect the pharmacokinetics (i.e., metabolism) or pharmacodynamics (i.e., mechanism of action) of an antidepressant medication may affect response and/or lead to side effects.[27–29] To address this issue, pharmacogenetic testing has been developed: this evaluates the genotypes of specific alleles and variants involved in pharmacokinetics and pharmacodynamics to determine which drugs an individual would be most likely to tolerate.[30–32] The usefulness of this testing method is currently a matter of debate, and preliminary studies have produced mixed results.[31,33] In a randomized controlled trial (RCT) of nearly 1400 participants with TRD, pharmacogenetic-guided treatment was compared to treatment as usual (TAU).[31] After 8 weeks, there was no difference between groups on symptom reduction scores but there was a significantly higher response rate among the guided-treatment group (26%) compared with the TAU group (20%).[31] Furthermore, when patients receiving a treatment incongruent with what was recommended by their pharmacogenetic test results were switched to a congruent treatment, they experienced a significantly greater response and lower side effect burden than those who remained on incongruent medication.[31] In a meta-analysis of over 1700 participants with MDD, similarly positive results were obtained: participants receiving pharmacogenetic-guided treatment were 1.7 times more likely to achieve remission than those receiving TAU.[33]

The relationship between comorbid medical disorders and TRD is complex. Individuals with untreated medical comorbidities may appear to have TRD, or these medical comorbidities may worsen depressive symptoms. For example, fibromyalgia, chronic fatigue syndrome, and irritable bowel syndrome are strongly linked with depressive symptoms.[34] To treat these individuals, medications specifically targeting the underlying medical conditions, often in combination with antidepressants, may be required. The most prevalent example of this is hypothyroidism: hypothyroidism is more common in individuals with TRD than in the general MDD population (52% vs. 8%–17%),[35] and treatment of hypothyroidism may lead to improvement in depressive symptoms.[36] Certain medical conditions may also prevent response to psychotropic medication, and treating these underlying conditions may improve response to standard antidepressant therapies. Diabetes, coronary artery disease, HIV infection, and cancer are all associated with a lower rate of antidepressant response.[37] Finally, commonly prescribed nonpsychiatric medications, such as glucocorticosteroids and antihypertensives, are associated with depressive symptoms and may prevent remission.[34]

Psychiatric comorbidities, such as substance abuse, obsessive-compulsive disorder, and post-traumatic stress disorder, may complicate diagnosis and treatment. Patients with these comorbidities may not achieve response with antidepressants only and other psychiatric medications or psychotherapy may need to be added.[25] Following the failure of two or more standard antidepressants, it is recommended that family physicians refer their patients to a psychiatrist. Psychiatrists are more likely to prescribe less conventionally used drugs such as TCAs or MAOIs, in addition to other combination and adjunctive strategies, and can address psychiatric comorbidities.

Etiopathology

Comorbid psychiatric disorders, such as anxiety, substance abuse, and personality disorders, may increase the risk of TRD.[34] The reason for this is twofold: first, comorbid psychiatric disorders are often associated with greater symptom severity[34] and second, treatment for multiple diagnoses is more complicated and these secondary diagnoses may not receive adequate treatment.[34]

Individuals who have experienced a greater number of stressful life events, such as immigration, death of a family member, interpersonal relationship problems, or financial stress, are at greater risk for treatment resistance.[38] In fact, one study (N = 137) identified some form of childhood adversity, such as trauma and bullying, in over 60% of a hospitalized group of patients with TRD.[39] In a systematic review of 50 treatment outcome studies in MDD, higher levels of neuroticism predicted a worse outcome.[40] Similarly, using the Temperament and Character Inventory[41] in a study comparing depressed individuals with and without TRD (N = 128), higher levels of neuroticism and lower levels of extraversion, openness to experiences, and conscientiousness were more prevalent in the TRD group compared with the non-TRD group and healthy controls.[42] Lower levels of openness in TRD may be related to lower levels of resilience, which may also be a risk factor for TRD.[42] Other clinical risk factors for TRD include older age, earlier age of MDD onset, longer duration of an MDE, more frequent and recurrent MDEs, melancholia, higher baseline severity of depressive illness, and suicidality.[43–45]

Dysregulation of neurotransmitters, including monoamines, γ-aminobutyric acid (GABA), and glutamate, is implicated in MDD and TRD. The monoamine hypothesis of depression focuses on reduced availability of monoamines, including serotonin, norepinephrine, and dopamine.[46–48] Over time, with advances in neuroimaging, it was found that dysfunction of the glutamatergic and GABA-mediated (-GABAergic) systems also play a critical role in MDD pathophysiology.[49,50] Glutamate dysregulation can cause long-term changes in brain areas and circuits implicated in mood regulation.[51] Furthermore, reduced concentrations and altered expression of GABA receptors are observed in MDD, and these reductions have been linked to cellular, behavioral, and cognitive changes found in depression.[52] Alterations in these neurotransmitter systems are particularly associated with TRD[44]; for example, greater deficits in GABA transmission are found in those with TRD compared with healthy controls and non-TRD individuals with MDD.[53,54]

Changes in neurotransmitter gene expression have also been implicated in TRD. A genomic study found that individuals with MDD carrying a polymorphism in the 5-HT_{1A} (serotonin) receptor, specifically *5-HT*$_{1A}$ *C1019G polymorphism GG genotype*, were over three times more likely to be treatment resistant.[55] Polymorphisms in the 5-HTT promoter region of the serotonin transporter gene have also been implicated in treatment nonresponse.[56] Many studies have found that individuals with MDD who have the short allele at the serotonin transporter polymorphism site that maps to the promoter region, known as 5-HTTLPR, are less likely to respond to SSRIs than those carrying the long allele,[57–59] but other studies have failed to replicate this association.[60,61] Finally, a functional polymorphism in the gene GRIN_{2B} that codes for the 2B subunit of the *N*-methyl-D-aspartate (NMDA) receptor, a major glutamate receptor, was associated with increased risk of treatment resistance in MDD.[62]

Brain-derived neurotrophic factor (BDNF), which is involved in neurodevelopment and neuroplasticity, has also been implicated in major depressive pathology, with individuals with MDD showing lower levels of peripheral BDNF than healthy controls.[44] The Val66Met polymorphism in BDNF was found to increase risk of TRD by over three times in one study[55]; however, this was not replicated in a separate investigation.[63]

With regard to brain structure, a smaller hippocampal volume has been identified as a risk factor for treatment nonresponse[64] and a larger hippocampal tail volume as a predictor of early and sustained response.[65] Furthermore, hyperactivity in the subgenual anterior cingulate cortex, left dorsomedial prefrontal cortex, putamen, pallidum, and amygdala has been implicated in TRD.[66] Other brain regions implicated in TRD include the entorhinal cortex, nucleus accumbens (NAc), medial forebrain bundle (MFB), and limbic-cortical-striatal-pallidal-thalamic circuit.[45]

Beyond abnormal neurotransmitters and brain circuit dysfunction, proinflammatory cytokines and immune system dysfunction have been of increasing interest in relation to MDD pathology.[45] TRD is associated with hypothalamic-pituitary-adrenal axis disturbance, proliferative T-cell activity, and overall activation of the inflammatory system.[45]

Treatment strategies

For TRD, monotherapy or polytherapy treatment options may be considered. Monotherapy, specifically switching to a different antidepressant, is usually the starting point, as this eliminates the risk of adverse drug interactions and places a lower financial burden on patients. However, when psychiatric medications are combined, they may have complementary pharmacodynamics that enhance antidepressant outcome. Furthermore, if a drug shows partial efficacy, it may be preferable to combine this drug with an adjunctive agent, rather than switch medications entirely. Monotherapy strategies are outlined in Table 3.3 and polytherapy strategies in Table 3.4. The level of evidence and recommendations for included medications are presented here, as well. Criteria for level of evidence and recommendations are outlined in Table 3.2.

Table 3.2 Criteria for level of evidence and recommendation.

Level of Evidence	
1	Meta-analysis with narrow confidence intervals *and/or* ≥2 RCTs with adequate sample size, ideally placebo-controlled
2	Meta-analysis with wide confidence intervals *and/or* ≥1 RCT with adequate sample size
3	RCT with small sample size, nonrandomized study, controlled prospective study, case series, *and/or* high-quality retrospective study
Recommendation	
1	Level 1 evidence with clinical support
2	Level 2 evidence with clinical support *or* level 1 evidence with clinical support but higher side effect burden than level 1 recommendation options
3	Level 3 evidence with clinical support *or* level 2 evidence with clinical support but higher side effect burden than level 2 recommendation options
Not recommended	Null findings in ≥1 RCT with adequate sample size *or* lack of data available to support recommendation and currently available evidence (e.g., small case series) shows poor efficacy

RCT, *randomized controlled trial*

Adapted from the Kennedy SH, Lam RW, McIntyre RS, et al. Canadian network for mood and anxiety treatments (CANMAT) 2016 clinical guidelines for the management of adults with major depressive disorder: section 3. pharmacological treatments. The Canadian Journal of Psychiatry. *2016;61(9):540–560. doi:10.1177/0706743716659417.*

Table 3.3 Antidepressant switching strategies and recommendations.

Antidepressant agent[a] (dosage)	Recommendation
Selective serotonin reuptake inhibitor (varies)	1
Serotonin-norepinephrine reuptake inhibitor (varies)[b]	
Agomelatine (25–50 mg/d)	
Bupropion (150–300 mg/d)	
Mianserin (60–120 mg/d)	
Mirtazapine (15–45 mg/d)	
Vortioxetine (10–20 mg/d)	
Tricyclic antidepressant (varies)	2
Moclobemide (300–600 mg/d)	
Quetiapine (150–400 mg/d)	
Selegiline transdermal (6–12 mg/d)	
Trazodone (150–300 mg/d)	
Vilazodone (20–40 mg/d)[c]	
Phenelzine (45–90 mg/d)	3
Reboxetine (8–10 mg/d)	
Tranylcypromine (20–60 mg/d)	

mg/d, *milligrams per day*
[a] *All level 1 evidence.*
[b] *Excluding levomilnacipran.*
[c] *Titrate from 10 mg*
Adapted from the Kennedy SH, Lam RW, McIntyre RS, et al. Canadian network for mood and anxiety treatments (CANMAT) 2016 clinical guidelines for the management of adults with major depressive disorder: section 3. pharmacological treatments. The Canadian Journal of Psychiatry. *2016;61(9):540–560. doi:10.1177/0706743716659417. only including antidepressants studied for treatment-resistant depression.*

Switching antidepressants

The switching strategy is typically the next step after poor tolerability or complete nonresponse to an initial antidepressant.[25] Clinicians may choose to switch to a different class of antidepressant (e.g., from an SSRI to an SNRI) or switch within the same class (e.g., citalopram to fluoxetine). While some evidence suggests that switching between classes and within a class is equally effective,[2,67,68] other investigators have reported superior outcomes when switching from an SSRI to an SNRI, specifically switching to venlafaxine.[25,69,70]

Although the vast majority of studies look at switching from an SSRI, as this is typically the first class of antidepressant prescribed, data show efficacy is comparable when switching between any antidepressant classes.[2,69] Aside from switching between SSRI and SNRI antidepressants, switching to mirtazapine, bupropion, or vortioxetine, or a TCA or MAOI, should also be considered. Decisions about which agent to select are usually based on specific symptom targets; for example, insomnia might favor mirtazapine[71] while anergia would suggest bupropion.[72] Overall, individuals switched from an SSRI to mirtazapine show remission rates comparable to

Table 3.4 Augmentation strategies for treatment-resistant depression, recommendations, and level of evidence.

Level of evidence	Adjunctive agent (dosage)[a]	Recommendation
1	Aripiprazole (2–15 mg/d)	1
	Risperidone (1–3 mg/d)	
	Quetiapine (150–300 mg/d)	
	Brexpiprazole (1–3 mg/d)	
	Olanzapine + fluoxetine (3 + 25 to 12 + 50 mg/d)	2
	Mirtazapine (30–60 mg/d)	
	IN esketamine (56–84 mg, *bis in 7 d*)	
	Pindolol	Not recommended (lack of efficacy)
	Buspirone	
	Dexmecamylamine	
2	Bupropion (150–450 mg/d)[b]	2
	Mianserin (30–60 mg/d)	
	Lithium (600–1200 mg/d)	
	T_3 (25–50 μg/d)	
	Modafinil (100–400 mg/d)	
	Benzodiazepines (varies)	3
	Lamotrigine (50–400 mg/d)	
	Pramipexole (0.25–2.25 mg/d)	
	Riluzole (50–100 mg bid)	
	Buprenorphine + samidorphan (2+2 –8+8 mg/d)	
	Methylphenidate	Not recommended (lack of efficacy)
	Lisdexamfetamine	
	Infliximab	
	Testosterone	
3	Reboxetine (2–8 mg/d)	3
	Ziprasidone (20–80 mg bid)	
	Amantadine (100–300 mg/d)	
	Topiramate (100–200 mg/d)	
	Celecoxib (200–400 mg/d or 100 –200 mg bid)	
	Minocycline (100–400 mg/d)	
	Scopolamine (4 μg/kg)[c]	
	Mecamylamine (5 mg/d or 5 mg bid)	
	Estrogen	Not recommended (lack of efficacy)
	Pregabalin	
	Gabapentin	
	Sirukumab	

Continued

Table 3.4 Augmentation strategies for treatment-resistant depression, recommendations, and level of evidence.—*cont'd*

Level of evidence	Adjunctive agent (dosage)[a]	Recommendation
Investigational	Tocilizumab	
	Psychedelics	

bid, *twice per day;* bis in 7 d, *twice per week;* IN, *intranasal;* mg/d, *milligrams per day;* μg/d, *micrograms per day;* μg/kg, *micrograms per kilogram*
[a] *Not all drugs available in North America.*
[b] *Extended release (bupropion XL) may be taken in one dose, immediate release should be taken in multiple doses of 150 mg throughout the day.*
[c] *Three infusions total, given 3–5 days apart*
Adapted from the Kennedy SH, Lam RW, McIntyre RS, et al. Canadian network for mood and anxiety treatments (CANMAT) 2016 clinical guidelines for the management of adults with major depressive disorder: section 3. pharmacological treatments. The Canadian Journal of Psychiatry. *2016;61(9):540–560. doi:10.1177/0706743716659417.*

Table 3.5 Neuromodulation treatments for treatment-resistant depression, recommendations, and level of evidence.

Level of evidence[a]	Neuromodulation	Recommendation
1	Repetitive transcranial magnetic stimulation	1
	Intermittent theta burst stimulation	
	Electroconvulsive therapy	2
2	Vagus nerve stimulation	3
3	Transcranial direct current stimulation	
	Deep brain stimulation	Investigational
	Magnetic seizure therapy	
	Trigeminal nerve stimulation	

[a] *Efficacy for acute treatment.*
Adapted from the Kennedy SH, Lam RW, McIntyre RS, et al. Canadian network for mood and anxiety treatments (CANMAT) 2016 clinical guidelines for the management of adults with major depressive disorder: section 3. pharmacological treatments. The Canadian Journal of Psychiatry. *2016;61(9):540–560. doi:10.1177/0706743716659417.*

those who switched to another SSRI,[73] although participants in one trial experienced faster response and remission when switched to mirtazapine versus an SSRI.[74] Elsewhere, significantly fewer participants achieved remission when switched to mirtazapine compared to venlafaxine[75] and experienced similar levels of remission when switched to mirtazapine or nortriptyline.[76] Individuals with TRD switched to bupropion experience comparable remission rates to those switched to an SSRI or SNRI,[2] and bupropion may be a good option for patients experiencing antidepressant-related sexual dysfunction, as this drug is associated with little to no sexual side effects.[77,78]

When comparing remission rates across studies, switching to vortioxetine led to numerically higher remission rates than switching to sertraline, venlafaxine, bupropion, or citalopram.[79,80] Like bupropion, vortioxetine may be a good switching option for patients experiencing treatment-emergent sexual dysfunction, as vortioxetine switching led to significantly greater improvements in sexual functioning than escitalopram.[81]

In most developed nations, TCAs have been largely replaced by newer antidepressants with better safety and tolerability profiles, but not superior efficacy.[82] Interestingly, melancholic patients in particular may benefit from switching to a TCA.[83] Owing to the increased toxicity in overdose and higher burden of adverse events associated with TCAs, it is recommended that they should only be prescribed when patients have failed several trials with newer antidepressants.[25]

There is limited evidence on the efficacy of switching to an MAOI for TRD, as the majority of these studies assess switching from a TCA. MAOIs were historically recommended for individuals with atypical features, but evidence to support this is also limited: MAOIs are more effective than TCAs, but not SSRIs, for the treatment of atypical depression.[84] MAOIs are typically recommended only after other drug classes have been ineffective because of the need for careful attention to food and drug interactions associated with MAOIs.[12,85]

Combination and adjunctive therapies

When an antidepressant shows good tolerability and partial efficacy, combining antidepressants may be preferable to switching. Combination strategies avoid drug discontinuation symptoms and cross-tapering,[86] and combining two antidepressants may lead to complementary neuropharmacologic mechanisms that enhance efficacy further than when taking either antidepressant alone.[86] Of course, clinicians must check for any contraindications or potential adverse interactions that may result from combining two different drugs. In the literature, combination therapy has been generally understudied, especially in TRD populations.[25,86] Another option for polytherapy treatment is augmentation; this involves the addition of a psychoactive medication that is not an antidepressant to a standard antidepressant. This strategy is generally used when the patient has tolerated the current antidepressant well and obtained partial response. For the sake of clarity, we will use the term "adjunctive" in this section to describe both combination and augmentation treatment strategies.

Two of the most frequently prescribed adjunctive antidepressants are bupropion and mirtazapine. However, the evidence to support bupropion, in combination with an SSRI, is weak.[87] Mirtazapine is often combined with SSRIs and SNRIs, based on the suggestion that this combination enhances monoaminergic neurotransmission, improves symptoms of insomnia, and counteracts gastrointestinal side effects of SSRIs/SNRIs.[88] The efficacy of this combination strategy is equivocal: when participants with TRD were prescribed mirtazapine plus venlafaxine, the remission rate was only 14% after an average of 8 weeks in the STAR*D study,[89] but the remission rate was 38% after 12 weeks in another large study.[90] Elsewhere, in a clinical trial comparing fluoxetine monotherapy to fluoxetine plus mirtazapine for MDD, but not TRD specifically, the combination was associated with a 52% remission rate, versus only 25% with fluoxetine alone.[91]

Two other adjunctive antidepressants that are approved in Europe, but not in North America, are mianserin and reboxetine. Both show limited efficacy: in two separate TRD populations, addition of mianserin to fluoxetine led to a significant decrease in depressive symptoms,[92] but there was no benefit in adding mianserin to sertraline.[93] Reboxetine shows the lowest efficacy of all major antidepressants, as shown in a meta-analysis,[82] but there may be some evidence for its efficacy in TRD specifically, with remission rates between 35% and 69% when reboxetine was added to an SSRI or SNRI in three midsize studies.[94–96]

Atypical antipsychotics

Atypical antipsychotics are the most studied class of non-antidepressant adjunctive agents for SSRIs and SNRIs.[12,16] In addition, because many of these agents have undergone global trials to obtain an indication as adjunctive agents when one or more traditional antidepressant(s) were not effective, both the size and quality of these databases are substantially greater than those of other adjunctive agents. Atypical antipsychotics are distinguished from "typical" antipsychotics by their action on both dopamine and serotonin receptors, acting in most instances as partial agonists or antagonists.[97]

In placebo controlled trials involving atypical antipsychotics, rates of remission are achieved at about twice the rate with adjunct atypical antipsychotics compared to placebo.[98–100] However, several atypical agents are associated with significant adverse events including weight gain, insulin resistance, dyslipidemia, and metabolic syndrome.[101] Extrapyramidal symptoms are also prevalent with certain atypical agents.[102] In fact, discontinuation rates due to adverse events are three times greater with atypical antipsychotics than with placebo.[98] Special caution should be taken when administering atypical antipsychotics to geriatric patients, especially individuals with dementia or other neurodegenerative diseases, as there is evidence to suggest atypical antipsychotics confer an increased risk of cerebrovascular events and death in these populations.[101,103]

As of 2020, only the specific combination of olanzapine and fluoxetine has FDA approval for TRD, and analyses support greater remission rates with this combination than fluoxetine alone.[104,105] The antidepressant benefits of adjunctive aripiprazole are well established: in multiple clinical trials and meta-analyses, aripiprazole performs significantly better than placebo in improving depressive symptoms for TRD, showing significantly higher response and remission rates.[100,106,107] Brexpiprazole is a relatively new atypical antipsychotic that, like aripiprazole, acts as a dopamine multifunctional agent with a low potential for inducing extrapyramidal symptoms.[108] However, brexpiprazole has a higher affinity for 5-HT_{1A} than D_2 receptors, whereas the opposite is true with aripiprazole.[108,109] At 2 and 3 mg/day, but not 1 mg/day, brexpiprazole adjunct treatment led to greater improvements in depressive symptoms and functioning than placebo after 6 weeks in participants with TRD (total N = 1853).[110–112] However, although brexpiprazole has a lower risk of akathisia than aripiprazole,[108] the incidence of akathisia increases with brexpiprazole dose, with rates of 14% at 3 mg.[110] Therefore it is not recommended to exceed 2 mg/day in depressed patients unless clinically necessary.[113]

Other atypical antipsychotics that have been investigated in TRD, with supportive evidence for their use when initial monotherapy is ineffective, include quetiapine, risperidone, and ziprasidone. Among these, quetiapine extended release currently demonstrates the most evidence for its efficacy in TRD.[100,106] Risperidone was also found to improve depressive symptoms more rapidly than placebo and led to higher remission and response rates[114–117]; this difference was not maintained in a maintenance trial.[117] Ziprasidone is the least studied among atypical agents for TRD.[118,119]

Olanzapine, quetiapine, and risperidone can lead to significant weight gain, with olanzapine being the most culpable.[101,120–122] The risk of weight gain is low for aripiprazole and brexpiprazole and negligible for ziprasidone.[120–124] However, ziprasidone has been associated with the potentially lethal cardiac conduction problem torsades de pointes and with "drug reaction with eosinophilia and systemic symptoms" (DRESS).[101,125] Furthermore, aripiprazole and brexpiprazole are frequently associated with elevated levels of akathisia.[100,110] Finally, risperidone may increase prolactin levels in users.[120]

Lithium

Lithium, in addition to its role as an antimanic and prophylactic agent in bipolar disorder, has been recognized as an adjunctive agent for TRD since the 1980s.[126] Currently, adjunct lithium is not prescribed as often as it once was, although older evidence suggests it may be effective in TRD.[127] In a 2015 meta-analysis, adjunct lithium led to significantly higher response rates than placebo in TRD but was poorly tolerated compared with other common augmentation strategies.[106] The majority of adjunct lithium trials have looked at augmentation to a TCA, making it difficult to determine the efficacy of lithium added to newer antidepressants. However, in the STAR*D study of lithium added to citalopram for TRD, the remission rate was 16%.[128]

Thyroid hormone (T_3)

Like lithium, T_3 was more popular in the past as an adjunctive agent for TRD; however, T_3 has been less studied than lithium. In a meta-analysis, T_3 was found to be significantly more effective than placebo in TRD and was better tolerated than other common adjunctive agents, such as quetiapine, olanzapine, and aripiprazole.[106] As with lithium, the majority of studies have assessed the efficacy of T_3 as an adjuvant to TCAs. However, when T_3 was combined with citalopram in STAR*D, 25% of patients with TRD remitted.[128]

Pindolol

Pindolol is a β-blocker and $5\text{-}HT_{1A}$ antagonist that is approved for the treatment of hypertension. The latter effect prompted some investigators to see if pindolol enhances the antidepressant properties of SSRIs.[129] However, a recent meta-analysis found pindolol was not more effective than placebo for TRD, with similar response and remission rates between the two groups.[106]

Anxiolytics

Individuals with TRD are more likely than those with non-treatment-resistant MDD to have comorbid anxiety disorders,[34] and anxious depression is a particularly difficult to treat subtype of depression.[25] Benzodiazepines may be most effective for treatment of depression with atypical features and accompanied by significant anxiety, mood reactivity, and/or hyperactivity.[130] A 2001 meta-analysis found that participants prescribed adjunct benzodiazepines were more likely to respond in 4 weeks than those treated with antidepressants alone (63% vs. 38%).[131] However, benzodiazepines have a high abuse potential, may be responsible for severe and prolonged discontinuation symptoms, and may lead to worsening of mood.[132] There is evidence that benzodiazepine use is associated with high levels of anhedonia, which could suggest benzodiazepines exacerbate anhedonia symptoms.[133] This is important for patients with TRD, as TRD is associated with higher levels of anhedonia than non-treatment-resistant MDD.[134] Owing to the lack of efficacy data beyond 4 weeks, benzodiazepines are not recommended for long-term use.[135,136] Furthermore, owing to the many risks associated with benzodiazepines, their use is recommended only in certain cases: specifically, MDD with anxiety, insomnia, catatonic features, and akathisia.[12,137] Adjunct buspirone, an anxiolytic not of the benzodiazepine class, has also been studied for TRD but shows no efficacy over placebo or other adjunctive strategies in several RCTs.[138–140]

Psychostimulants and stimulantlike drugs

Psychostimulants are typically prescribed for the treatment of attention-deficit/hyperactivity disorder and act primarily on the dopaminergic system to promote wakefulness and cognitive enhancement.[141] Although it is common in clinical practice to prescribe adjunct psychostimulants for TRD, the evidence to support this strategy is limited.[25,141] Any benefits observed from adjunct psychostimulant therapy in MDD tend to be rapid, but short-lived.[142] Two meta-analyses found adjunctive therapy with methylphenidate or lisdexamfetamine was not more effective than placebo for improving depressive symptoms or achieving a clinical response in TRD.[106,143] Caution must be taken when prescribing psychostimulants because of their potential for abuse and the possible worsening of anxiety, irritability, and/or insomnia symptoms.[144]

Modafinil, a stimulant-like wakefulness-promoting agent indicated for the treatment of narcolepsy and obstructive sleep apnea, is a norepinephrine and dopamine modulator that is often prescribed in practice for TRD. The effects on these neurotransmitter levels are less pronounced than with psychostimulants, decreasing the risk of abuse and side effects.[145] In a pooled analysis of two RCTs involving TRD participants who partially responded to SSRIs, adjunct modafinil rapidly and significantly improved depressive symptoms and overall clinical condition compared with placebo after 1 week, and this improvement was maintained at week 8.[146] Modafinil also rapidly improves common secondary symptoms of depression, including fatigue, daytime sleepiness, and some aspects of cognitive dysfunction.[147]

Dopaminergic agents

Two dopaminergic agents used primarily for treatment of Parkinson's disease, pramipexole and amantadine, have been assessed as potential adjunctive therapies in depression. While adjunctive pramipexole was superior to placebo in decreasing depressive symptoms in TRD, there were no statistical differences in response or remission rates between these groups.[148] Caution should be taken when combining pramipexole and escitalopram: this combination was associated with a substantially higher dropout rate than when either treatment was prescribed alone (69% vs. 15%), likely due to more severe side effects.[149]

Amantadine, in addition to its dopaminergic and noradrenergic actions, acts as an NMDA receptor antagonist. In a systematic review of preclinical and clinical studies of the antidepressant effects of amantadine, the authors concluded that amantadine might potentiate the effects of other antidepressants.[150] The first trial of amantadine for TRD was conducted in 1971, in which the authors reported 67% of those receiving adjunctive amantadine improved in terms of depressive symptoms versus only 25% in the placebo group.[151] Subsequently, individuals with TRD were treated with open-label imipramine plus amantadine or imipramine only for 6 weeks, and 20% of participants prescribed amantadine experienced a response versus none in the imipramine-only group.[152]

Anticonvulsants

Lamotrigine, a voltage-gated sodium channel blocker, is currently approved for use in epilepsy and bipolar disorder. In a systematic review and meta-analysis, adjunctive lamotrigine was found to have a significant, but modest, antidepressant effect in participants with TRD.[153] This meta-analysis also suggested that lamotrigine may be more effective for severely ill patients (HAM-D $\geq$24): the number needed to treat with lamotrigine was nine, but decreased to four when only severely ill patients were considered.[153]

Riluzole, a glutamate antagonist, is used as an anticonvulsant and for the treatment of amyotrophic lateral sclerosis. Even though this drug showed preliminary efficacy in small, open-label trials as an adjunctive agent,[154,155] subsequent larger-scale RCTs and meta-analyses did not confirm the efficacy of riluzole in treating TRD.[156,157] However, riluzole may have delayed antidepressant effects. In the largest adjunct riluzole RCT to date (N = 104), no improvement in depressive symptoms was found after 4–8 weeks of riluzole compared to placebo,[156] but 24% of initial nonresponders to riluzole responded at week 20 in the open-label extension study.[158]

Topiramate is used for the treatment of epilepsy and migraine. Its mechanism of action is relatively unknown, but it may act as a sodium channel blocker, GABA agonist, and/or a GABA-receptor antagonist.[159] A single RCT in TRD found that topiramate added to an SSRI significantly reduced depression scores after 8 weeks compared with placebo, and 60% of topiramate-treated participants met response criteria versus 0% of placebo.[160]

There is a paucity of data exploring gabapentin and pregabalin for TRD. In one RCT, pregabalin did not affect response or remission rates when added to escitalopram.[161] Gabapentin has only been explored in a chart review.[162] Without further TRD-targeted trials, there is insufficient evidence to support the use of either agent in TRD.[163,164]

Anti-inflammatory agents

A growing body of evidence implicates inflammatory processes in the etiology of MDD. Individuals with MDD present with higher levels of proinflammatory cytokines, such as interleukin 1, interleukin 6, and tumor necrosis factor α (TNF-α), as well as elevated C-reactive protein levels.[165,166] In fact, a meta-analysis found that higher levels of baseline inflammation contributed to treatment resistance in depressed individuals.[167] Therefore several anti-inflammatory agents have been investigated for their potential antidepressant properties.

Adjunctive therapy with celecoxib, a cyclooxygenase 2 (COX-2) inhibitor, has shown significant efficacy over placebo in reducing the severity of depression.[168–171] Unfortunately, no studies to date have examined the efficacy of adjunct celecoxib in TRD populations specifically.

Minocycline, an antibiotic used to treat various bacterial infections, rheumatoid arthritis, and acne, also reduces or inhibits interleukin 1B, TNF-α, prostaglandin E_2, and COX-2.[172,173] Positive effects of adjunctive minocycline in an open-label trial in geriatric TRD (clinicaltrials.gov, NCT01659320) were replicated in an RCT for midlife MDD.[174]

Infliximab is an anti-inflammatory agent that acts as a TNF-α blocker. An RCT found that baseline TNF-α concentrations were significantly higher in infliximab responders versus nonresponders.[175] However, a meta-analysis of adjunct infliximab for TRD found no significant antidepressant effect of infliximab.[176]

Tocilizumab and sirukumab, both interleukin 6 receptor inhibitors used to treat rheumatoid arthritis, have also been evaluated for putative antidepressant properties. Both tociluzumab[177–179] and sirukumab[180] reduced the severity of depressive symptoms in rheumatoid arthritis trials. One clinical trial to date has assessed the effects of sirukumab on depressive symptoms in MDD and found no benefit for the drug compared to placebo (clinicaltrials.gov, NCT02473289). As of April 2020, there are ongoing clinical trials assessing tocilizumab in depression,[181] one of which is looking at TRD specifically (clinicaltrails.gov, NCT02660528).

Sex steroids

In 2001, Schneider and colleagues[182] found that among women who were 60 years or older with MDD and taking sertraline, those who were also on estrogen replacement therapy (ERT) had significantly greater improvement in overall clinical condition and quality of life compared with those not on ERT. Two subsequent small studies in perimenopausal women with TRD found ERT significantly reduced depression scores[183,184]; however, another study found no effect of adjunctive therapy.[185] Overall, the results for adjunct estrogen for MDD and TRD are inconclusive and limited.[186] Adjunct testosterone therapy in men with TRD has also been studied,[187–189] most commonly for men with comorbid hypogonadism.[188,189] Of the three RCTs identified, all found that although mood did improve with testosterone, the improvement was not significantly different compared to placebo.[187–189]

Nicotinic antagonists

Stimulation of nicotinic acetylcholine receptor systems is hypothesized to have antidepressant effects, based on the mood-enhancing properties of nicotine.[190] This hypothesis prompted a clinical trial program to evaluate mecamylamine, a nicotinic receptor antagonist, as a potential adjunctive agent for TRD. While an early placebo-controlled RCT showed a significant reduction in depressive symptoms in the mecamylamine plus SRRI group compared with placebo after 8 weeks,[191] in a much larger phase III trial of dexmecamylamine (the S-enantiomer of mecamylamine) added to an SSRI or SNRI for TRD, there was no difference between active and placebo groups.[192]

Muscarinic cholinergic receptor antagonists

There is evidence to support a cholinergic-MDD hypersensitivity hypothesis in at least a subgroup of individuals with TRD.[193] To test this hypothesis, scopolamine, an antimuscarinic anticholinergic agent used to treat motion sickness and postoperative nausea and vomiting, was evaluated in a small placebo-controlled crossover trial as a potential treatment for TRD. Three infusions of scopolamine or placebo were administered over 9–15 days.[194] Participants experienced a significantly greater response to scopolamine versus placebo, with a significant improvement in depressive symptoms found after just one intravenous (IV) infusion.[194] Although this has not been pursued further, several reviews include the antidepressant effects of scopolamine in psychiatric disorders (not specifically TRD) and conclude that scopolamine is a potentially effective antidepressant agent with rapid onset of action, eliciting an antidepressant effect in as little as 3 days and lasting for at least 2 weeks.[193,195]

Opioids

Certain opiate receptor subtypes are known to modulate serotonin regulation in the brain,[196] and brain areas rich in opiate receptors receive projections from other regions associated with mood regulation, including the amygdala, frontal cortex, and locus coeruleus.[197] In fact, several studies provide support for the hypothesis that deficiency of endogenous opioids may underlie the pathogenesis of MDD.[198] However, synthetic opioids are associated with significant addictive potential, so extreme caution must be taken when prescribing these drugs as antidepressants. In two placebo-controlled RCTs, the synthetic opioid buprenorphine was tested in a TRD sample and combined with samidorphan in hopes of counteracting the addictive potential of buprenorphine.[199,200] When participants with TRD (N = 32 and 142) were given buprenorphine + samidorphan or placebo (7 days or 4 weeks), as adjunctive therapy to an SSRI or SNRI, a significant decrease in depressive symptoms, compared with placebo, was observed in the investigational group.[199,200] However, this difference was only observed when buprenorphine and samidorphan were given in equal parts and not when participants were given more buprenorphine than samidorphan (at an 8:1 ratio)[199] or higher doses (8 vs. 2 mg buprenorphine).[200] The incidence of treatment-emergent adverse events with this drug combination was high (86%), most commonly related to gastrointestinal and neurologic symptoms.[200]

Nutraceuticals

While data are limited compared to pharmacologic treatments, some studies suggest antidepressant effects of natural products. Folate, *S*-adenosylmethionine (SAMe), omega-3 fatty acids, *N*-acetylcysteine (NAC), and acetyl-L-carnitine (ALC) are known to play important roles in proper brain function.[201–204] Biological levels of SAMe, omega-3, and ALC are decreased in individuals with MDD compared with healthy volunteers,[201,205–207] as are levels of glutathione, which are enhanced by cysteine.[208]

In a midsize placebo-controlled RCT of adjunctive L-methylfolate for TRD, 15 mg/day, but not 7.5 mg/day, adjunct L-methylfolate was associated with a significantly greater antidepressant response compared with placebo after 60 days.[201] Furthermore, another midsize RCT found 800 mg of adjunct SAMe prescribed for 8 weeks was superior to placebo in a TRD sample.[209] Finally, when adjunct omega-3 fatty acid was compared to adjunct lithium, participants with TRD prescribed omega-3 experienced a significant decrease in depressive symptoms, which was comparable to lithium.[210] No placebo-controlled trials to date have studied the efficacy of adjunct NAC or ALC for improving symptoms in TRD. However, NAC was not superior to placebo in a large MDD sample after 12 weeks but was superior for improving functional capacity.[211] Furthermore, a meta-analysis found ALC was significantly more effective than placebo in improving depressive symptoms across a variety of psychiatric disorders.[212] Before a conclusion regarding the efficacy of these supplements for depressive symptoms can be made, large-scale placebo-controlled studies are needed.

Neuromodulation

Neuromodulation is "the process of inhibition, stimulation, modification, regulation or therapeutic alteration of activity, electrically or chemically, in the central, peripheral, or autonomic nervous systems".[213] Neuromodulation techniques have been employed in disorders such as chronic pain, movement disorders, epilepsy, and psychiatric disorders.[213] For TRD, neuromodulation techniques may be invasive or noninvasive. Noninvasive devices are designed to stimulate neural tissue, typically through the scalp. Invasive techniques involve the implantation of the neuromodulatory device directly on the region that is to be stimulated and a pacemaker in the chest to control the frequency of stimulation. The different types of neuromodulation used in TRD are discussed in the following and outlined in Table 3.5.

Electroconvulsive therapy

ECT serves as the prototypic method of neurostimulation in depression. Although the exact mechanism of action for ECT is unknown, the dominating hypothesis is that induction of a seizure causes changes in neurotransmitters, neuroplasticity, and functional connectivity in the brain.[214–218] Between 6 and 15 treatments, given two to three times per week, are typically required to achieve response and/or remission; treatment more than three times per week is associated with greater cognitive side effects.[214]

ECT is one of the most effective treatments for MDD and the most effective treatment for severe TRD.[214,219,220] Approximately 50% of patients with TRD respond to ECT[221]; however, in a meta-analysis, more than half of individuals relapsed within the first year.[222] It is unclear whether adjunct antidepressant medication is beneficial during a course of ECT: it does not appear to improve immediate response rates but may reduce likelihood of relapse.[222,223] Importantly, taking antidepressants during the course of ECT may increase the risk of memory deterioration associated with ECT.[223] After a successful course of ECT, maintenance strategies include subsequent prescription of antidepressant medication, psychotherapy, or continuation ECT (C-ECT).[214,222] With pharmacotherapy or C-ECT, relapse rates are reduced by around 50%.[222] The data on relapse rates with maintenance psychotherapy is limited; one RCT found relapse rates were lower when cognitive behavior therapy (CBT) was combined with C-ECT compared with C-ECT alone.[224]

Even though both clinicians and patients commonly associate ECT with cognitive impairment, there is no compelling evidence of long-term cognitive impairments after ECT and any cognitive symptoms appear to be transient, typically returning to pretreatment levels in a few days to weeks.[225,226] However, bitemporal electrode placement has been consistently associated with more negative cognitive effects than bifrontal or unilateral placement.[227]

Magnetic seizure therapy

Magnetic seizure therapy (MST) uses electromagnetic induction to produce an electric field in the brain and ultimately leads to a more focused tonic-clonic seizure, targeting only the prefrontal cortex.[228] This therapy has a lower burden of side effects compared with ECT.[229] Guidelines for the delivery of MST are similar to those for ECT, with treatment two to three times a week for approximately 12 treatments total.[214] Although MST is newer and much less studied than ECT, data thus far shows MST is superior to sham treatment and associated with response and relapse rates comparable to ECT.[228,230,231] MST appears to be more effective at higher stimulation frequencies, with 100 Hz achieving the highest remission rates.[232] While MST and ECT have been directly compared in a small pilot study, showing comparable efficacy,[231] results from a large-scale direct comparison are still awaited (ClinicalTrials.gov, NCT03191058).

Repetitive transcranial magnetic stimulation

Via an induction coil placed against the scalp, repetitive transcranial magnetic stimulation (rTMS) uses powerful focused magnetic field pulses to induce electric currents in brain tissue.[233] rTMS is most commonly applied over the dorsolateral prefrontal cortex (DLPFC), with high-frequency rTMS applied to the left side of DLPFC and low-frequency rTMS applied to the right side of DLPFC.[234] Although the exact mechanism of action of rTMS is currently unknown, it is hypothesized that rTMS increases activity in the left side of DLPFC, which is thought to be underactive in MDD, and decreases activity in the right side of DLPFC, which is proposed to be overactive in MDD.[235] In addition, rTMS induces changes in mRNA expression of genes implicated in depression, as well as in blood flow and activity in brain regions associated with depression pathology.[236]

Greater treatment effects from rTMS are achieved at higher frequencies (20 Hz) and lower intensities (100% or less), after 15–30 treatments and when used in combination with antidepressant medication.[220,237] Bilateral and unilateral rTMS show comparable efficacy.[237] Results from multiple large-scale studies and meta-analyses of rTMS in participants with TRD yielded response rates of 29%–46% and remission rates between 16% and 31%.[220,235,237–240] However, antidepressant effects may not be lasting: in several studies, there was no difference between rTMS and sham treatment groups with regard to depression scores and response rates after 3 or 4 months.[241–243] Overall, few studies have assessed the long-term efficacy of rTMS and more research is required before a definitive conclusion can be made. Although the role of rTMS has yet to be established in TRD, its noninvasive nature and relatively low side effect burden justifies its use with pharmacotherapy or other treatments.

Theta burst stimulation

In order to improve the clinical utility of rTMS, theta burst stimulation (TBS) was developed.[218,244] TBS requires a shorter stimulation time and lower intensity of stimulation than rTMS, which in turn produces longer-lasting modulation of cortical excitability.[245] TBS can be applied continuously (cTBS), intermittently (iTBS), or as a combination of both and is typically applied to the prefrontal cortex when used for MDD.[245] In inferiority trials comparing iTBS to rTMS, iTBS was noninferior to rTMS after 4–6 weeks of treatment.[246,247] In a 2015 meta-analysis of TBS for MDD, response rates were significantly greater with TBS (36%) than with sham treatment (18%), but there was no difference in remission rates (19% vs. 11%).[245] Specifically, left-side DLPFC iTBS and bilateral DLPFC cTBS/iTBS performed significantly better than sham, but right-side DLPFC cTBS did not.[245] These findings were corroborated in several subsequent TBS trials.[248,249] Interestingly, in long-term follow-up studies, response rates continued to increase 2 weeks (10%) and 14 weeks (18%) later.[248,250]

Deep brain stimulation

Deep brain stimulation (DBS) delivers electric impulses to the brain via electrodes implanted in key brain regions. A pacemaker implanted in the chest sets the stimulation frequency. DBS was originally developed for the treatment of movement disorders, such as essential tremor and Parkinson's disease,[251–254] but it has been recognized as an experimental intervention for TRD since 2005.[255] While the majority of trials in TRD have targeted the subcallosal cingulate gyrus (SCC) Brodmann area 24/25, other regions have been tested for electrode placement, including the ventral internal capsule/ventral striatum (VC/VS), NAc, inferior thalamic peduncle, habenula, MFB), bed nucleus of stria terminalis (BST), and anterior limb of internal capsule (ALIC).[256,257]

The majority of these trials are open label and relatively small. In the seminal report, six participants were given SCC DBS: after 1 month, two were responders and after another month, all but one participant met response criteria, which was maintained by four participants at 6-month follow-up.[255] An important aspect of this trial was the integration of positron emission tomographic imaging to identify regional changes in blood flow following treatment.[255] An extension of the original cohort to 20 participants reported a response rate at 1 year of 60%,[258] and these improvements were maintained 3–6 years after surgery.[258,259] Since these initial studies, many more SCC DBS studies for TRD have been conducted, including several sham versus active stimulation RCTs, and the reported efficacy has varied greatly, with response rates between 20% and 82% and remission rates between 18% and 80% after 6–12 months of stimulation.[260–265] Furthermore, in the largest SCC DBS study to date (N = 90), no difference in response was found between active and sham treatment after 4–6 months.[262] About 30% of participants experienced one or more serious adverse events, with at least a quarter of these cases related directly to the study device or surgery.[262]

SCC DBS is by far the most studied form of DBS for TRD; however, other electrode placements have shown positive results in smaller, pilot studies. Initial impressive results were found with DBS of the superolateral MFB: among 16 people given this treatment, all achieved response after 2 months, with 50% responding in just 1 week.[266] Antidepressant effects have also been found with DBS applied to the NAc, ITP, habenula, and BST, but these were in case series or studies with very small sample sizes (N = 5–11).[267–272] One study of 25 participants with TRD reported a 40% response rate with ALIC DBS, with response maintained at 2 years.[273,274] However, 40% of participants dropped out before the study was completed because of lack of efficacy and/or unstable psychiatric status.[273] Finally, VC/VS DBS was not superior to sham treatment in 29 people with TRD, and this treatment modality has been associated with several adverse events, including hypomania, increased depression and suicidality, perioperative pain, and syncope.[275]

When DBS is turned off, patients may experience a rapid recurrence of depressive symptoms.[276] Therefore continuous administration of DBS is important for effective treatment, and caution should be taken when turning off or explanting a patient's device.

Vagus nerve stimulation

Vagus nerve stimulation (VNS) involves the stimulation of the left cervical vagus nerve using an implantable device that elicits intermittent electric signals to this nerve region. Again, the exact mechanism of action of VNS is unknown, but it is hypothesized that VNS excites several brain regions directly or indirectly connected to the vagus nerve, including the midbrain, hypothalamus, amygdala, BST, insular cortex, and limbic forebrain areas.[277] Furthermore, preclinical studies have found VNS affects the serotonergic and noradrenergic neurotransmitter systems and BDNF receptors.[277] VNS is typically administered adjunctively to antidepressant medication.[278] Individuals on medium- (0.5−1 mA current, 250 μs pulse width) to high-stimulation (1.25−1.5 mA, 250 μs) settings may be more likely to maintain response than those on low-stimulation settings (0.25 mA, 130 μs)[279] Overall, in 13 studies evaluating VNS as an augmentation to standard of care, the pooled response rate was 24% at 6 months, 39% at 1 year, and 53% at 2 years, as determined in a systematic review and meta-analysis.[278] In a longitudinal observational study in the United States, when individuals with TRD were treated with VNS ($N = 494$) or TAU ($N = 301$), the response rate for VNS after 5 years was 68%, which was significantly greater than that in the TAU arm (41%).[280]

Trigeminal nerve stimulation

Trigeminal nerve stimulation (TNS) is a noninvasive technique involving electric stimulation of the trigeminal nerve. The trigeminal nerve directly projects onto the nucleus tractus solitarius, locus coeruleus, reticular activating system, and thalamic structures and thereby indirectly acts on sensory, limbic, and other cortical and subcortical structures in the brain.[281] Although TNS was originally developed for the control of seizures in treatment-resistant epilepsy, investigators also observed antidepressant effects.[282]

More recently, TNS has been tested for its antidepressant effects in mood disorders, including TRD.[281,283] In the initial pilot study of TNS for TRD, 55% of participants responded after 8 weeks of treatment.[283] Another pilot study found a 100% response rate after just 10 days; only one participant did not meet remission criteria.[284] To date, only one RCT of TNS for TRD has been completed: after 4 weeks of TNS therapy or sham treatment, active TNS was superior to sham in decreasing depressive symptoms ($N = 40$).[285] This difference remained significant at 30-day follow up.[285]

Transcranial direct current stimulation

Transcranial direct current stimulation (tDCS) delivers a continuous low-amplitude electric current to specific regions of the cortex via noninvasive electrodes placed on the scalp. Two types of stimulation are used: anodal stimulation excites neuronal activity, while cathodal stimulation inhibits activity.[286] tDCS is thought to induce neuroplasticity by modulating synaptic strength in the cortex in a similar manner to long-term potentiation/depression.[286] The main region of stimulation for tDCS is the DLPFC. tDCS is prescribed for TRD as monotherapy or augmentation to standard antidepressants.[287] In three randomized, sham-controlled clinical trials (N = 22–25), there was no difference between active tDCS and sham stimulation at any time point (1–3 weeks) with regard to depressive severity improvement or response rates.[288–290] Interestingly, in an open-label study of tDCS (N = 18), no significant improvements in depressive symptoms were observed until week 6, suggesting tDCS may have a delayed treatment effect.[291] No RCTs followed participants for this length of time, so emergence of late-onset symptom improvements could not be determined.

Psychotherapy

Psychotherapy is recommended for TRD in addition to TAU, i.e., pharmacologic treatment and/or neurostimulation.[292] In a recent systematic review of TRD, the most common forms of psychotherapy were CBT, cognitive behavior analysis system of psychotherapy (CBASP), mindfulness-based cognitive therapy (MBCT), and interpersonal psychotherapy (IPT).[292] When psychotherapy and TAU are given in conjunction, the antidepressant effect is greater than either alone.[292] MBCT shows the greatest efficacy (pooled effect size, g = 0.55), followed by CBASP (g = 0.42), IPT (g = 0.33), and CBT (g = 0.26).[292] In a meta-analysis looking at the long-term efficacy of CBT and MBCT, significant improvements in depressive symptoms were maintained for at least 12 months.[293] Group therapy, as opposed to individual treatment, is associated with greater reductions in depressive severity.[292]

Psychedelics

Evidence that psychedelics may have a therapeutic role in the management of TRD has emerged over several years leading up to 2020. Psychedelic therapy involves administering an oral dose of a psychedelic drug to induce a psychedelic, or mystical, experience lasting 4–9 hours, and it should be accompanied by some form of psychotherapy.[294] The hope is that the patient will work through difficult feelings and emotions while in this mystical state, which is hypothesized to lead to lasting antidepressant and anxiolytic effects.[295] Among candidate drugs for psychedelic therapy, psilocybin, lysergic acid diethylamide (LSD), 3,4-methylenedioxymethamphetamine (MDMA), and ayahuasca are under investigation. Psychedelics are thought to exert antidepressant effects by acting as serotonin receptor agonists, thereby increasing glutamate-dependent activity and BDNF levels in prefrontal brain regions and suppressing inflammatory processes of the immune system.[296] This, in turn, is hypothesized to enhance neuroplasticity and have neuroprotective effects.[296] When psychedelics are taken in an enriched context (i.e., a comforting environment with adjunct psychotherapy), cognitive biases may be reversed during this state of neuroplasticity.[297] For optimal safety and efficacy, the participant should be contained in a safe and calming environment and monitored by a therapist for the duration of their experience.[294]

Psilocybin

During the resurgence of clinical and scientific interest in psilocybin, known colloquially as "magic mushrooms" or "shrooms," it has been studied for its antidepressant effects in TRD in two related clinical trials. In 2016 and 2018, Carhart-Harris and colleagues[297,298] recruited 12 and 20 participants with TRD, respectively, prescribing two doses of psilocybin (10 and 25 mg) 7 days apart. Psychologic support was provided before, during, and after the psychedelic experience.[297] In both studies, there were significant improvements in depression, anxiety, and anhedonia after 1 week and this was maintained 3 months later.[297,298] The degree of depression improvement was correlated with self-reported quality of acute psychedelic experience.[297] Reported side effects were well-tolerated and transient, including confusion, headache, nausea, and paranoia.[297,298]

Lysergic acid diethylamide

LSD, a semisynthetic psychedelic drug, was studied for its antidepressant properties in several small-scale studies before its legal prohibition in 1967.[299] In a systematic review, data on 423 individuals across 19 studies were combined and a 79% "improvement" rate in depressive symptoms was found.[300] However, there were significant methodological issues with the papers included in this review, including anecdotal evidence, inadequate assessment procedures, lack of controls, and a very large range of LSD doses across studies (20–1500 μg).[300] Furthermore, neither the definition for depression "improvement" was provided, nor was level of treatment resistance specified.[300]

3,4-Methylenedioxymethamphetamine

MDMA has not been studied for TRD or MDD, but several studies have included depression measures when evaluating the effects of MDMA in post-traumatic stress disorder. In a pooled analysis of results from four phase II RCTs, improvement in depression symptoms was numerically greater for the MDMA group (75–125 mg) than the placebo/control group after 1–2 months, with a trend toward significance ($P = .053$).[301] MDMA was well tolerated.[301]

Ayahuasca

Ayahuasca is a psychedelic brew, or tea, containing dimethyltryptamine, used traditionally by aboriginal populations for ritualistic and/or medicinal purposes.[294] Longitudinal studies in ritual users show an association between use and reduced incidence of mental health problems.[294,302] In two open-label clinical trials ($N = 6$ and 17), when participants with TRD were administered a single dose of ayahuasca, significant improvements in depressive symptoms were observed after just 1 day, and maintained at 21-day follow-up.[303,304] In a subsequent placebo-controlled trial (N = 29), ayahuasca was significantly better than placebo in improving depressive symptoms 7 days after a single dose.[305] In all three clinical trials, no long-term adverse events were observed, but vomiting did occur in 47%−57% of those receiving ayahuasca.[303−305]

N-methyl-D-aspartate receptor antagonists: ketamine and esketamine

Ketamine, delivered IV, elicits rapid reductions in depressive symptoms, with patients with TRD experiencing mood improvements in as little as 40 min.[306] In TRD, ketamine is most commonly administered IV at 0.5 mg/kg given over 40−45 min.[307] Multiple studies have investigated the effects of a single ketamine infusion in TRD: response rates between 64% and 80% were achieved after just 24 h.[306,308−310] However, the majority of participants relapsed in these trials, with mean time to relapse varying from 13 to 22 days.[310,311] Subsequently, studies have assessed the efficacy of repeated IV ketamine infusions. When six ketamine infusions were given over 12 days, response rates on day 12 ranged from 59% to 96%.[312−315] However, on follow-up, the majority of participants relapsed, with a mean time to relapse of 16−19 days.[312−314] When maintenance infusions were delivered weekly for 4 weeks, this protocol was able to prevent relapse in this time frame.[315] Interestingly, when six infusions were given over a longer period, i.e., 3 weeks, response rates were lower, between 23% and 58%, and relapse rates remained high.[316−318] It is important to note that ketamine is still investigational, and these studies have all been limited by small sample sizes. Currently, a large-scale clinical trial is underway in Canada comparing the efficacy of ECT and repeated-dose IV ketamine infusions given over several months for TRD (ClinicalTrials.gov, NCT03674671).

Intranasal (IN), subcutaneous, and intramuscular delivery systems for ketamine have also been studied in TRD. Although 50 mg IN ketamine led to a rapid reduction in depressive symptoms after just 24 h,[319] this route of administration has poor tolerability when patients are asked to self-administer.[320] When a small study compared IV, intramuscular, and subcutaneous routes of administration, all routes had comparable antidepressant efficacy, but the subcutaneous route was associated with the least amount of side effects.[321] Finally, in a small sample of geriatric TRD, repeated subcutaneous ketamine injections led to a 50% remission rate.[322]

The European Medicines Agency and the FDA approved a delivery system with IN esketamine, the S-enantiomer of ketamine, for TRD in 2019. IN esketamine is typically prescribed as an adjuvant to standard oral antidepressant medications, administered twice weekly at 56 or 84 mg. In two systematic reviews, IN esketamine displayed rapid and significant reductions in depressive symptoms lasting at least 28 days when given at doses of 56 mg.[323,324] However, one large-scale study (N = 346) found no significant improvement in depressive symptoms after 4 weeks of 84 mg IN esketamine treatment.[325] There are also considerable concerns over the cost-effectiveness of esketamine and if its efficacy outweighs the risks,[326,327] as esketamine is associated with tolerability issues such as dizziness, dysgeusia, somnolence, dissociation, suicidal thoughts and behaviors, and increased heart rate and blood pressure.[323] In fact, the National Institute for Health and Care Excellence in the United Kingdom recently issued a statement against the use of IN esketamine for depression.[327]

Conclusion

TRD is a complex psychiatric diagnosis with varying degrees of severity. TRD may be difficult to diagnose, as many confounding factors, such as poor treatment adherence and comorbid medical or psychiatric conditions, may lead to misdiagnosis. Consideration of comorbidities, life stressors, personality traits, and genetic and biological factors can help in determining the risk of developing TRD. Until the diagnostic sensitivity can identify subtypes within MDD with distinct therapeutic targets, current strategies should be followed sequentially. Currently, the standard of care for a patient who presents as treatment resistant starts with optimization strategies, followed by switching or combination of standard antidepressant medications. Adjunctive strategies are the logical next step. In more severe cases of TRD, neuromodulation may be effective. ECT is currently the most successful treatment for TRD, but other neuromodulation options such as rTMS and DBS may also be employed. Finally, a number of exciting new treatments are currently under investigation, namely, psychedelics and ketamine, and have shown remarkable efficacy in early pilot trials. As new treatment emerges, and as we gain a better understanding of the neurobiology of TRD, we may now begin to wonder: will today's TRD be tomorrow's inadequately treated depression?

Box 3.1 Defining treatment resistant depression

- Treatment-resistant depression is broadly defined as a failure to achieve an adequate response to an adequate trial of an antidepressant therapy.
- There is no consensus among researchers and/or clinicians on what constitutes an "adequate" response or an "adequate" antidepressant trial.
- Adequate response is either defined as a relatively asymptomatic state or a $\geq$50% decrease in depressive symptoms, measured using a validated depression rating scale (e.g., Hamilton Depression Rating Scale).
- An adequate trial must include an approved antidepressant medication taken at a stable dose for a period deemed adequate to achieve a therapeutic effect (e.g., 8 weeks).

Box 3.2 Prevalence and diagnosis

- About 50% of patients do not respond to their first antidepressant treatment.
- The likelihood of achieving remission decreases as the number of failed antidepressant trials increases.

- When diagnosing patients with treatment-resistant depression (TRD), first consider other factors such as poor drug adherence, pharmokinetic factors, medical conditions, and psychiatric comorbidities.
- Risk factors for TRD include psychiatric and medical comorbidities, psychosocial factors, dysregulation of certain neurotransmitters, and immune dysfunction.

References

1. Nemeroff CB. Prevalence and management of treatment-resistant depression. *The Journal of Clinical Psychiatry.* 2007;68(suppl.8):17–25.
2. Rush AJ, Trivedi MH, Wisniewski SR, et al. Acute and longer-term outcomes in depressed outpatients requiring one or several treatment steps: a STAR*D report. *The American Journal of Psychiatry.* 2006;163(11):1905–1917. https://doi.org/10.1176/ajp.2006.163.11.1905.
3. Petersen T, Papakostas GI, Mahal Y, et al. Psychosocial functioning in patients with treatment resistant depression. *European Psychiatry.* 2004;19(4):196–201. https://doi.org/10.1016/j.eurpsy.2003.11.006.
4. American Psychiatric Association. *DSM*. Vol. 5. 2013, 10.1176/appi.books.9780890425596.744053.
5. World Health Organization. *International Classification of Disease 11th Revision*; 2018. https://icd.who.int/browse11/l-m/en.
6. Brown S, Rittenbach K, Cheung S, McKean G, MacMaster FP, Clement F. Current and common definitions of treatment-resistant depression: findings from a systematic review and qualitative interviews. *The Canadian Journal of Psychiatry.* 2019;64(6):380–387. https://doi.org/10.1177/0706743719828965.
7. Sackeim HA. The definition and meaning of treatment-resistant depression. *The Journal of Clinical Psychiatry.* 2001;62(suppl.16):10–17. Suppl 1 http://www.ncbi.nlm.nih.gov/pubmed/11480879.
8. Hamilton M. A rating scale for depression. *Journal of Neurology Neurosurgery, and Psychiatry.* 1960;23:56–62. https://doi.org/10.1136/jnnp.23.1.56.
9. Montgomery SA, Asberg M. A new depression scale designed to be sensitive to change. *British Journal of Psychiatry.* 1979;134(4):382–389. https://doi.org/10.1192/bjp.134.4.382.
10. Gaynes BN, Lux L, Gartlehner G, et al. Defining treatment-resistant depression. *Depression and Anxiety.* 2020;37(2):134–145. https://doi.org/10.1002/da.22968.
11. Fava M. Diagnosis and definition of treatment-resistant depression. *Biological Psychiatry.* 2003;53(8):649–659. https://doi.org/10.1016/S0006-3223(03)00231-2.
12. Kennedy SH, Lam RW, McIntyre RS, et al. Canadian network for mood and anxiety treatments (CANMAT) 2016 clinical guidelines for the management of adults with major depressive disorder: section 3. pharmacological treatments. *The Canadian Journal of Psychiatry.* 2016;61(9):540–560. https://doi.org/10.1177/0706743716659417.
13. Sackeim HA, Prudic J, Devanand DP, Decina P, Kerr B, Malitz S. The impact of medication resistance and continuation pharmacotherapy on relapse following response to

electroconvulsive therapy in major depression. *Journal of Clinical Psychopharmacology.* 1990;10(2):96–104. https://doi.org/10.1097/00004714-199004000-00004.
14. Souery D, Amsterdam J, De Montigny C, et al. Treatment resistant depression: methodological overview and operational criteria. *European Neuropsychopharmacology.* 1999; 9(1–2):83–91. https://doi.org/10.1016/S0924-977X(98)00004-2.
15. Fekadu A, Wooderson S, Donaldson C, et al. A multidimensional tool to quantify treatment resistance in depression: the maudsley staging method. *The Journal of Clinical Psychiatry.* 2009;70(2):177–184. https://doi.org/10.4088/JCP.08m04309.
16. McIntyre RS, Filteau MJ, Martin L, et al. Treatment-resistant depression: definitions, review of the evidence, and algorithmic approach. *Journal of Affective Disorders.* 2014;156:1–7. https://doi.org/10.1016/j.jad.2013.10.043.
17. Berlim MT, Turecki G. What is the meaning of treatment resistant/refractory major depression (TRD)?. A systematic review of current randomized trials. *European Neuropsychopharmacology.* 2007;17(11):696–707. https://doi.org/10.1016/j.euroneuro.2007.03.009.
18. Ruhé HG, van Rooijen G, Spijker J, Peeters FPML, Schene AH. Staging methods for treatment resistant depression. A systematic review. *Journal of Affective Disorders.* 2012;137(1–3):35–45. https://doi.org/10.1016/j.jad.2011.02.020.
19. Fawcett J. Progress in treatment-resistant and treatment-refractory depression: we still have a long way to go. *Psychiatric Annals.* 1994;24(5):214–216. https://doi.org/10.3928/0048-5713-19940501-05.
20. Oquendo MA, Malone KM, Ellis SP, Sackeim HA, Mann JJ. Inadequacy of antidepressant treatment for patients with major depression who are at risk for suicidal behavior. *The American Journal of Psychiatry.* 1999;156(2):190–194. https://doi.org/10.1176/ajp.156.2.190.
21. Thase ME, Rush AJ. When at first you don't succeed: sequential strategies for antidepressant nonresponders. *Journal of Clinical Psychiatry.* 1997;58(Suppl 1):23–29.
22. Gartlehner G, Hansen RA, Morgan LC, et al. Comparative benefits and harms of second-generation antidepressants for treating major depressive disorder. *Annals of Internal Medicine.* 2011;155(11):772. https://doi.org/10.7326/0003-4819-155-11-201112060-00009.
23. Rizvi SJ, Grima E, Tan M, et al. Treatment-resistant depression in primary care across Canada. *The Canadian Journal of Psychiatry.* 2014;59(7):349–357. https://doi.org/10.1177/070674371405900702.
24. Rush AJ. STAR*D: what have we learned? *The American Journal of Psychiatry.* 2007; 164(2):201–204. https://doi.org/10.1176/ajp.2007.164.2.201.
25. Ionescu DF, Rosenbaum JF, Alpert JE. Pharmacological approaches to the challenge of treatment-resistant depression. *Dialogues in Clinical Neuroscience.* 2015;17(2): 111–126.
26. Sansone RA, Sansone LA. Antidepressant adherence: are patients taking their medications? *Innovations in Clinical Neuroscience.* 2012;9(5–6):41–46.
27. Sim SC, Risinger C, Dahl M-L, et al. A common novel CYP2C19 gene variant causes ultrarapid drug metabolism relevant for the drug response to proton pump inhibitors and antidepressants. *Clinical Pharmacology and Therapeutics.* 2006;79(1):103–113. https://doi.org/10.1016/j.clpt.2005.10.002.
28. Murphy GM, Kremer C, Rodrigues HE, Schatzberg AF. Pharmacogenetics of antidepressant medication intolerance. *The American Journal of Psychiatry.* 2003;160(10): 1830–1835. https://doi.org/10.1176/appi.ajp.160.10.1830.

29. Tansey KE, Guipponi M, Hu X, et al. Contribution of common genetic variants to antidepressant response. *Biological Psychiatry.* 2013;73(7):679–682. https://doi.org/10.1016/j.biopsych.2012.10.030.
30. Hall-Flavin DK, Winner JG, Allen JD, et al. Using a pharmacogenomic algorithm to guide the treatment of depression. *Translation Psychiatry.* 2012;2(10):e172. https://doi.org/10.1038/tp.2012.99.
31. Greden JF, Parikh SV, Rothschild AJ, et al. Impact of pharmacogenomics on clinical outcomes in major depressive disorder in the GUIDED trial: a large, patient- and rater-blinded, randomized, controlled study. *Journal of Psychiatric Research.* 2019;111:59–67. https://doi.org/10.1016/j.jpsychires.2019.01.003.
32. Spear BB, Heath-Chiozzi M, Huff J. Clinical application of pharmacogenetics. *Trends in Molecular Medicine.* 2001;7(5):201–204. https://doi.org/10.1016/S1471-4914(01)01986-4.
33. Bousman CA, Arandjelovic K, Mancuso SG, Eyre HA, Dunlop BW. Pharmacogenetic tests and depressive symptom remission: a meta-analysis of randomized controlled trials. *Pharmacogenomics.* 2019;20(1):37–47. https://doi.org/10.2217/pgs-2018-0142.
34. Kornstein SG, Schneider RK. Clinical features of treatment-resistant depression. *The Journal of Clinical Psychiatry.* 2001;62(suppl6):18–25.
35. Howland RH. Thyroid dysfunction in refractory depression: implications for pathophysiology and treatment. *The Journal of Clinical Psychiatry.* 1993;54(2):47–54.
36. Gold MS. Hypothyroidism and depression. Evidence from complete thyroid function evaluation. *JAMA Journal of American Medical Association.* 1981;245(19):1919–1922. https://doi.org/10.1001/jama.245.19.1919.
37. Popkin MK. The outcome of antidepressant use in the medically ill. *Archives of General Psychiatry.* 1985;42(12):1160. https://doi.org/10.1001/archpsyc.1985.01790350034007.
38. Amital D, Fostick L, Silberman A, Beckman M, Spivak B. Serious life events among resistant and non-resistant MDD patients. *Journal of Affective Disorders.* 2008;110(3):260–264. https://doi.org/10.1016/j.jad.2008.01.006.
39. Tunnard C, Rane LJ, Wooderson SC, et al. The impact of childhood adversity on suicidality and clinical course in treatment-resistant depression. *Journal of Affective Disorders.* 2014;152–154(1):122–130. https://doi.org/10.1016/j.jad.2013.06.037.
40. Mulder RT. Personality pathology and treatment outcome in major depression: a review. *The American Journal of Psychiatry.* 2002;159(3):359–371. https://doi.org/10.1176/appi.ajp.159.3.359.
41. Cloninger CR, Przybeck TR, Svrakic DM, Wetzel RD. *The Temperament and Character Inventory (TCI): A Guide to its Development and Use.* 1994.
42. Takahashi M, Shirayama Y, Muneoka K, Suzuki M, Sato K, Hashimoto K. Low openness on the revised NEO personality inventory as a risk factor for treatment-resistant depression. *PloS One.* 2013;8(9):1–7. https://doi.org/10.1371/journal.pone.0071964.
43. De Carlo V, Calati R, Serretti A. Socio-demographic and clinical predictors of non-response/non-remission in treatment resistant depressed patients: a systematic review. *Psychiatry Research.* 2016;240:421–430. https://doi.org/10.1016/j.psychres.2016.04.034.
44. Bennabi D, Aouizerate B, El-Hage W, et al. Risk factors for treatment resistance in unipolar depression: a systematic review. *Journal of Affective Disorders.* 2015;171:137–141. https://doi.org/10.1016/j.jad.2014.09.020.

45. Murphy JA, Sarris J, Byrne GJ. A review of the conceptualisation and risk factors associated with treatment-resistant depression. *Depression Research and Treatment*. 2017; 2017. https://doi.org/10.1155/2017/4176825.
46. Bunney Jr WE, Davis JM. Norepinephrine in depressive reactions. *Archives of General Psychiatry*. 1965;13(6):483–494. https://doi.org/10.1001/archpsyc.1965.01730060001001.
47. Schildkraut JJ. The catecholamine hypothesis of affective disorders: a review of supporting evidence. *The American Journal of Psychiatry*. 1965;122(5):509–522. https://doi.org/10.1176/ajp.122.5.509.
48. Belujon P, Grace AA. Dopamine system dysregulation in major depressive disorders. *The International Journal of Neuropsychopharmacology*. 2017;20(12):1036–1046. https://doi.org/10.1093/ijnp/pyx056.
49. Niciu MJ, Ionescu DF, Richards EM, Zarate CA. Glutamate and its receptors in the pathophysiology and treatment of major depressive disorder. *Journal of Neural Transmission*. 2013;121(8):907–924. https://doi.org/10.1007/s00702-013-1130-x.
50. Choudary PV, Molnar M, Evans SJ, et al. Altered cortical glutamatergic and GABAergic signal transmission with glial involvement in depression. *Proceedings of the National Academy of Sciences of the United States of America*. 2005;102(43): 15653–15658. https://doi.org/10.1073/pnas.0507901102.
51. Sanacora G, Treccani G, Popoli M. Towards a glutamate hypothesis of depression. *Neuropharmacology*. 2012;62(1):63–77. https://doi.org/10.1016/j.neuropharm.2011.07.036.
52. Luscher B, Shen Q, Sahir N. The GABAergic deficit hypothesis of major depressive disorder. *Molecular Psychiatry*. 2011;16(4):383–406. https://doi.org/10.1038/mp.2010.120.
53. Price RB, Shungu DC, Mao X, et al. Amino acid neurotransmitters assessed by proton magnetic resonance spectroscopy: relationship to treatment resistance in major depressive disorder. *Biological Psychiatry*. 2009;65(9):792–800. https://doi.org/10.1016/j.biopsych.2008.10.025.
54. Levinson AJ, Fitzgerald PB, Favalli G, Blumberger DM, Daigle M, Daskalakis ZJ. Evidence of cortical inhibitory deficits in major depressive disorder. *Biological Psychiatry*. 2010;67(5):458–464. https://doi.org/10.1016/j.biopsych.2009.09.025.
55. Anttila S, Huuhka K, Huuhka M, et al. Interaction between 5-HT1A and BDNF genotypes increases the risk of treatment-resistant depression. *Journal of Neural Transmission*. 2007;114(8):1065–1068. https://doi.org/10.1007/s00702-007-0705-9.
56. Bonvicini C, Minelli A, Scassellati C, et al. Serotonin transporter gene polymorphisms and treatment-resistant depression. *Progress in Neuro-Psychopharmacology and Biological Psychiatry*. 2010;34(6):934–939. https://doi.org/10.1016/j.pnpbp.2010.04.020.
57. Coplan JD, Gopinath S, Abdallah CG, Berry BR. A neurobiological hypothesis of treatment-resistant depression-mechanisms for selective serotonin reuptake inhibitor non-efficacy. *Frontiers in Behavioral Neuroscience*. 2014;8(MAY):1–16. https://doi.org/10.3389/fnbeh.2014.00189.
58. Huezo-Diaz P, Uher R, Smith R, et al. Moderation of antidepressant response by the serotonin transporter gene. *British Journal of Psychiatry*. 2009;195(1):30–38. https://doi.org/10.1192/bjp.bp.108.062521.
59. Gressier F, Bouaziz E, Verstuyft C, Hardy P, Becquemont L, Corruble E. 5-HTTLPR modulates antidepressant efficacy in depressed women. *Psychiatric Genetics*. 2009; 19(4):195–200. https://doi.org/10.1097/YPG.0b013e32832cef0d.
60. Maron E, Tammiste A, Kallassalu K, et al. Serotonin transporter promoter region polymorphisms do not influence treatment response to escitalopram in patients with major

depression. *European Neuropsychopharmacology.* 2009;19(6):451–456. https://doi.org/10.1016/j.euroneuro.2009.01.010.

61. Lewis G, Mulligan J, Wiles N, et al. Polymorphism of the 5-HT transporter and response to antidepressants: randomised controlled trial. *British Journal of Psychiatry.* 2011; 198(6):464–471. https://doi.org/10.1192/bjp.bp.110.082727.
62. Zhang C, Li Z, Wu Z, et al. A study of N-methyl-D-aspartate receptor gene (GRIN2B) variants as predictors of treatment-resistant major depression. *Psychopharmacology.* 2014;231(4):685–693. https://doi.org/10.1007/s00213-013-3297-0.
63. Li Z, Zhang Y, Wang Z, et al. The role of BDNF, NTRK2 gene and their interaction in development of treatment-resistant depression: data from multicenter, prospective, longitudinal clinic practice. *Journal of Psychiatric Research.* 2013;47(1):8–14. https://doi.org/10.1016/j.jpsychires.2012.10.003.
64. Fonseka TM, MacQueen GM, Kennedy SH. Neuroimaging biomarkers as predictors of treatment outcome in Major Depressive Disorder. *Journal of Affective Disorders.* 2018; 233:21–35. https://doi.org/10.1016/j.jad.2017.10.049.
65. Nogovitsyn N, Muller M, Souza R, et al. Hippocampal tail volume as a predictive biomarker of antidepressant treatment outcomes in patients with major depressive disorder: a CAN-BIND report. *Neuropsychopharmacology.* 2020;45(2):283–291. https://doi.org/10.1038/s41386-019-0542-1.
66. Duhameau B, Ferré J-C, Jannin P, et al. Chronic and treatment-resistant depression: a study using arterial spin labeling perfusion MRI at 3Tesla. *Psychiatry Research: Neuroimaging.* 2010;182(2):111–116. https://doi.org/10.1016/j.pscychresns.2010.01.009.
67. Souery D, Serretti A, Calati R, et al. Switching antidepressant class does not improve response or remission in treatment-resistant depression. *Journal of Clinical Psychopharmacology.* 2011;31(4):512–516. https://doi.org/10.1097/JCP.0b013e3182228619.
68. Souery D, Serretti A, Calati R, et al. Citalopram versus desipramine in treatment resistant depression: effect of continuation or switching strategies. A randomized open study. *The World Journal of Biological Psychiatry.* 2011;12(5):364–375. https://doi.org/10.3109/15622975.2011.590225.
69. Ruhé HG, Huyser J, Swinkels JA, Schene AH. Switching antidepressants after a first selective serotonin reuptake inhibitor in major depressive disorder. *The Journal of Clinical Psychiatry.* 2006;67(12):1836–1855. https://doi.org/10.4088/JCP.v67n1203.
70. Papakostas GI, Fava M, Thase ME. Treatment of SSRI-resistant depression: a meta-analysis comparing within- versus across-class switches. *Biological Psychiatry.* 2008; 63(7):699–704. https://doi.org/10.1016/j.biopsych.2007.08.010.
71. Thase ME. Antidepressant treatment of the depressed patient with insomnia. *The Journal of Clinical Psychiatry.* 1999;60(Suppl 17):28–31.
72. Jefferson JW, Rush AJ, Nelson JC, et al. Extended-release bupropion for patients with major depressive disorder presenting with symptoms of reduced energy, pleasure, and interest: findings from a randomized, double-blind, placebo-controlled study. *The Journal of Clinical Psychiatry.* 2006;67(6):865–873. https://doi.org/10.4088/JCP.v67n0602.
73. Fang Y, Yuan C, Xu Y, et al. Comparisons of the efficacy and tolerability of extended-release venlafaxine, mirtazapine, and paroxetine in treatment-resistant depression. *Journal of Clinical Psychopharmacology.* 2010;30(4):357–364. https://doi.org/10.1097/JCP.0b013e3181e7784f.
74. Connolly KR, Thase ME. If at first you don't succeed: a review of the evidence for an-

tidepressant augmentation, combination and switching strategies. *Drugs*. 2011;71(1): 43–64. https://doi.org/10.2165/11587620-000000000-00000.

75. Baldomero EB, Ubago JG, Cercós CL, Ruiloba JV, Calvo CG, López RP. Venlafaxine extended release versus conventional antidepressants in the remission of depressive disorders after previous antidepressant failure: ARGOS study. *Depression and Anxiety*. 2005;22(2):68–76. https://doi.org/10.1002/da.20080.
76. Fava M, Rush AJ, Wisniewski SR, et al. A comparison of mirtazapine and nortriptyline following two consecutive failed medication treatments for depressed outpatients: a STAR*D report. *The American Journal of Psychiatry*. 2006;163(7):1161–1172. https://doi.org/10.1176/ajp.2006.163.7.1161.
77. Modell JG, Katholi CR, Modell JD, DePalma RL. Comparative sexual side effects of bupropion, fluoxetine, paroxetine, and sertraline. *Clinical Pharmacology and Therapeutics*. 1997;61(4):478–487. https://doi.org/10.1016/S0009-9236(97)90198-3.
78. Montejo AL, Montejo L, Navarro-Cremades F. Sexual side-effects of antidepressant and antipsychotic drugs. *Current Opinion in Psychiatry*. 2015;28(6):418–423. https://doi.org/10.1097/YCO.0000000000000198.
79. Thase ME, Danchenko N, Brignone M, et al. Comparative evaluation of vortioxetine as a switch therapy in patients with major depressive disorder. *European Neuropsychopharmacology*. 2017;27(8):773–781. https://doi.org/10.1016/j.euroneuro.2017.05.009.
80. Brignone M, Diamand F, Painchault C, Takyar S. Efficacy and tolerability of switching therapy to vortioxetine versus other antidepressants in patients with major depressive disorder. *Current Medical Research and Opinion*. 2016;32(2):351–366. https://doi.org/10.1185/03007995.2015.1128404.
81. Jacobsen PL, Mahableshwarkar AR, Chen Y, Chrones L, Clayton AH. Effect of vortioxetine vs. Escitalopram on sexual functioning in adults with well-treated major depressive disorder experiencing SSRI-induced sexual dysfunction. *The Journal of Sexual Medicine*. 2015;12(10):2036–2048. https://doi.org/10.1111/jsm.12980.
82. Cipriani A, Furukawa TA, Salanti G, et al. Comparative efficacy and acceptability of 21 antidepressant drugs for the acute treatment of adults with major depressive disorder: a systematic review and network meta-analysis. *Lancet*. 2018;391(10128):1357–1366. https://doi.org/10.1016/S0140-6736(17)32802-7.
83. Perry PJ. Pharmacotherapy for major depression with melancholic features: relative efficacy of tricyclic versus selective serotonin reuptake inhibitor antidepressants. *Journal of Affective Disorders*. 1996;39(1):1–6. https://doi.org/10.1016/0165-0327(96)00014-6.
84. Henkel V, Mergl R, Allgaier AK, Kohnen R, Möller HJ, Hegerl U. Treatment of depression with atypical features: a meta-analytic approach. *Psychiatry Research*. 2006; 141(1):89–101. https://doi.org/10.1016/j.psychres.2005.07.012.
85. McCabe BJ. Dietary tyramine and other pressor amines in MAOI regimens: a review. *Journal of the American Dietetic Association*. 1986;86(8):1059–1064.
86. Carvalho AF, Berk M, Hyphantis TN, McIntyre RS. The integrative management of treatment-resistant depression: a comprehensive review and perspectives. *Psychotherapy and Psychosomatics*. 2014;83(2):70–88. https://doi.org/10.1159/000357500.
87. Patel K, Allen S, Haque MN, Angelescu I, Baumeister D, Tracy DK. Bupropion: a systematic review and meta-analysis of effectiveness as an antidepressant. *Therapeutic Advances in Psychopharmacology*. 2016;6(2):99–144. https://doi.org/10.1177/2045125316629071.

88. Stahl S M, Lee-Zimmerman C, Cartwright S. Ann Morrissette D. Serotonergic drugs for depression and beyond. *Current Drug Targets*. 2013;14(5):578–585. https://doi.org/10.2174/1389450111314050007.
89. McGrath PJ, Stewart JW, Fava M, et al. Tranylcypromine versus venlafaxine plus mirtazapine following three failed antidepressant medication trials for depression: a STAR*D report. *The American Journal of Psychiatry*. 2006;163(9):1531–1541. https://doi.org/10.1176/ajp.2006.163.9.1531.
90. Rush AJ, Trivedi MH, Stewart JW, et al. Combining medications to enhance depression outcomes (CO-MED): acute and long-term outcomes of a single-blind randomized study. *The American Journal of Psychiatry*. 2011;168(7):689–701. https://doi.org/10.1176/appi.ajp.2011.10111645.
91. Blier P, Ward HE, Tremblay P, Laberge L, Hébert C, Bergeron R. Combination of antidepressant medications from treatment initiation for major depressive disorder: a double-blind randomized study. *The American Journal of Psychiatry*. 2010;167(3):281–288. https://doi.org/10.1176/appi.ajp.2009.09020186.
92. Ferreri M, Lavergne F, Berlin I, Payan C, Puech AJ. Benefits from mianserin augmentation of flouxetine in patients with major depression non-responders to fluoxetine alone. *Acta Psychiatrica Scandinavica*. 2001;103(1):66–72. https://doi.org/10.1034/j.1600-0447.2001.00148.x.
93. Licht RW, Qvitzau S. Treatment strategies in patients with major depression not responding to first-line sertraline treatment: a randomised study of extended duration of treatment, dose increase or mianserin augmentation. *Psychopharmacology*. 2002;161(2):143–151. https://doi.org/10.1007/s00213-002-0999-0.
94. Seguí J, López-Muñoz F, Álamo C, Camarasa X, García-García P, Pardo A. Effects of adjunctive reboxetine in patients with duloxetine-resistant depression: a 12-week prospective study. *Journal of Psychopharmacology*. 2010;24(8):1201–1207. https://doi.org/10.1177/0269881109102641.
95. Rubio G, San L, López-Muñoz F, Alamo C. Reboxetine adjunct for partial or nonresponders to antidepressant treatment. *Journal of Affective Disorders*. 2004;81(1):67–72. https://doi.org/10.1016/j.jad.2003.08.001.
96. Lòpez-Muñoz F, Álamo C, Rubio G, García-García P, Pardo A. Reboxetine combination in treatment-resistant depression to selective serotonin reuptake inhibitors. *Pharmacopsychiatry*. 2007;40(1):14–19. https://doi.org/10.1055/s-2007-958523.
97. Seeman P. Atypical antipsychotics: mechanism of action. *The Canadian Journal of Psychiatry*. 2002;47(1):29–40. https://doi.org/10.1177/070674370204700106.
98. Papakostas GI, Shelton RC, Smith J, Fava M. Augmentation of antidepressants with atypical antipsychotic medications for treatment-resistant major depressive disorder: a meta-analysis. *The Journal of Clinical Psychiatry*. 2007;68(6):826–831. https://doi.org/10.4088/JCP.v68n0602.
99. Nelson JC, Papakostas GI. Atypical antipsychotic augmentation in major depressive disorder: a meta-analysis of placebo-controlled randomized trials. *The American Journal of Psychiatry*. 2009;166(9):980–991. https://doi.org/10.1176/appi.ajp.2009.09030312.
100. Spielmans GI, Berman MI, Linardatos E, Rosenlicht NZ, Perry A, Tsai AC. Adjunctive atypical antipsychotic treatment for major depressive disorder: a meta-analysis of depression, quality of life, and safety outcomes. *PLoS Medicine*. 2013;10(3):

e1001403. https://doi.org/10.1371/journal.pmed.1001403.

101. Orsolini L, Tomasetti C, Valchera A, et al. An update of safety of clinically used atypical antipsychotics. *Expert Opinion on Drug Safety*. 2016;15(10):1329–1347. https://doi.org/10.1080/14740338.2016.1201475.
102. Divac N, Prostran M, Jakovcevski I, Cerovac N. Second-generation antipsychotics and extrapyramidal adverse effects. *Biomed Research International*. 2014;2014:1–6. https://doi.org/10.1155/2014/656370.
103. MacQueen GM, Frey BN, Ismail Z, et al. Canadian network for mood and anxiety treatments (CANMAT) 2016 clinical guidelines for the management of adults with major depressive disorder. Section 6: special populations: youth, women, and the elderly. *The Canadian Journal of Psychiatry*. 2016;61(9):588–603. https://doi.org/10.1177/0706743716659276.
104. Tamayo JM, Pica-Ruiz Y, Ruiz I. Olanzapine and fluoxetine combined as therapy for treatment-resistant depression: a systematic review. *Archivos de Neurociencias*. 2015;20(1):3–19. https://doi.org/10.2147/NDT.S127453.
105. Luan S, Wan H, Wang S, Li H, Zhang B. Efficacy and safety of olanzapine/fluoxetine combination in the treatment of treatmentresistant depression: a meta-analysis of randomized controlled trials. *Neuropsychiatric Disease and Treatment*. 2017;13:609–620. https://doi.org/10.2147/NDT.S127453.
106. Zhou X, Ravindran AV, Qin B, et al. Comparative efficacy, acceptability, and tolerability of augmentation agents in treatment-resistant depression: systematic review and network meta-analysis. *The Journal of Clinical Psychiatry*. 2015;76(4):e487–e498. https://doi.org/10.4088/JCP.14r09204.
107. Kennedy SH, Lam RW, Rotzinger S, et al. Symptomatic and functional outcomes and early prediction of response to escitalopram monotherapy and sequential adjunctive aripiprazole therapy in patients with major depressive disorder: a CAN-BIND-1 report. *The Journal of Clinical Psychiatry*. 2019;80(2):18m12202. https://doi.org/10.4088/JCP.18m12202.
108. Al Shirawi MI, Edgar NE, Kennedy SH. Brexpiprazole in the treatment of major depressive disorder. *Clinical Medicine Insights Therapeutics January*. 2017. https://doi.org/10.1177/1179559X17731801.
109. Maeda K, Sugino H, Akazawa H, et al. Brexpiprazole I: in vitro and in vivo characterization of a novel serotonin-dopamine activity modulator. *Journal of Pharmacology Experimental Therapeutics*. 2014;350(3):589–604. https://doi.org/10.1124/jpet.114.213793.
110. Thase ME, Zhang P, Weiss C, Meehan SR, Hobart M. Efficacy and safety of brexpiprazole as adjunctive treatment in major depressive disorder: overview of four short-term studies. *Expert Opinion in Pharmacotherapy*. 2019;20(15):1907–1916. https://doi.org/10.1080/14656566.2019.1638913.
111. Thase ME, Youakim JM, Skuban A, et al. Adjunctive brexpiprazole 1 and 3 mg for patients with major depressive disorder following inadequate response to antidepressants: a phase 3, randomized, double-blind study. *The Journal of Clinical Psychiatry*. 2015;76(9):1232–1240. https://doi.org/10.4088/JCP.14m09689.
112. Thase ME, Youakim JM, Skuban A, et al. Efficacy and safety of adjunctive brexpiprazole 2 mg in major depressive disorder: a phase 3, randomized, placebo-controlled study in patients with inadequate response to antidepressants. *The Journal of Clinical Psychiatry*. 2015;76(9):1224–1231. https://doi.org/10.4088/JCP.14m09688.
113. Otsuka America Pharmaceutical Inc. Rexulti (Brexpiprazole) Prescribing Information. *Rockville*. 2018.

114. Rapaport MH, Gharabawi GM, Canuso CM, et al. Effects of risperidone augmentation in patients with treatment-resistant depression: results of open-label treatment followed by double-blind continuation. *Neuropsychopharmacology*. 2006;31(11):2505–2513. https://doi.org/10.1038/sj.npp.1301113.
115. Keitner GI, Garlow SJ, Ryan CE, et al. A randomized, placebo-controlled trial of risperidone augmentation for patients with difficult-to-treat unipolar, non-psychotic major depression. *Journal of Psychiatric Research*. 2009;43(3):205–214. https://doi.org/10.1016/j.jpsychires.2008.05.003.
116. Mahmoud RA, Pandina GJ, Turkoz I, et al. Risperidone for treatment-refractory major depressive disorder. *Annals of Internal Medicine*. 2007;147(9):593–602. https://doi.org/10.7326/0003-4819-147-9-200711060-00003.
117. Reeves H, Batra S, May RS, Zhang R, Dahl DC, Li X. Efficacy of risperidone augmentation to antidepressants in the management of suicidality in major depressive disorder: a randomized, double-blind, placebo-controlled pilot study. *The Journal of Clinical Psychiatry*. 2008;69(8):1228–1236. https://doi.org/10.4088/JCP.v69n0805.
118. Papakostas GI, Fava M, Baer L, et al. Ziprasidone augmentation of escitalopram for major depressive disorder: efficacy results from a randomized, double-blind, placebo-controlled study. *The American Journal of Psychiatry*. 2015;172(12):1251–1258. https://doi.org/10.1176/appi.ajp.2015.14101251.
119. Dunner DL, Amsterdam JD, Shelton RC, Loebel A, Romano SJ. Efficacy and tolerability of adjunctive ziprasidone in treatment-resistant depression: a randomized, open-label, pilot study. *The Journal of Clinical Psychiatry*. 2007;68(7):1071–1077. https://doi.org/10.4088/jcp.v68n0714.
120. Leucht S, Cipriani A, Spineli L, et al. Comparative efficacy and tolerability of 15 antipsychotic drugs in schizophrenia: a multiple-treatments meta-analysis. *Lancet*. 2013; 382(9896):951–962. https://doi.org/10.1016/S0140-6736(13)60733-3.
121. Dayabandara M, Hanwella R, Ratnatunga S, Seneviratne S, Suraweera C, de Silva V. Antipsychotic-associated weight gain: management strategies and impact on treatment adherence. *Neuropsychiatric Disease and Treatment*. 2017;13:2231–2241. https://doi.org/10.2147/NDT.S113099.
122. Bak M, Fransen A, Janssen J, van Os J, Drukker M. Almost all antipsychotics result in weight gain: a meta-analysis. *PLoS One*. 2014;9(4):e94112. https://doi.org/10.1371/journal.pone.0094112.
123. Weiss C, Weiller E, Baker RA, et al. The effects of brexpiprazole and aripiprazole on body weight as monotherapy in patients with schizophrenia and as adjunctive treatment in patients with major depressive disorder. *International Clinical Psychopharmacology*. 2018;33(5):255–260. https://doi.org/10.1097/YIC.0000000000000226.
124. Das S, Barnwal P, Winston AB, Mondal S, Saha I. Brexpiprazole: so far so good. *Therapeutic Advances in Psychopharmacology*. 2016;6(1):39–54. https://doi.org/10.1177/2045125315614739.
125. U.S. Food and Drug Administration. *Drug Safety and Availability - FDA Drug Safety Communication: FDA Reporting Mental Health Drug Ziprasidone (Geodon) Associated with Rare but Potentially Fatal Skin Reactions*; 2014. https://www.fda.gov/drugs/drug-safety-and-availability/fda-drug-safety-communication-fda-reporting-mental-health-drug-ziprasidone-geodon-associated-rare.
126. Dé Montigny C, Grunberg F, Mayer A, Deschenes J-P. Lithium induces rapid relief of depression in tricyclic antidepressant drug non-responders. *British Journal of Psychiatry*. 1981;138(3):252–256. https://doi.org/10.1192/bjp.138.3.252.

127. Crossley NA, Bauer M. Acceleration and augmentation of antidepressants with lithium for depressive disorders. *The Journal of Clinical Psychiatry*. 2007;68(06):935–940. https://doi.org/10.4088/JCP.v68n0617.
128. Nierenberg AA, Fava M, Trivedi MH, et al. A comparison of lithium and T3 augmentation following two failed medication treatments for depression: a STAR*D report. *The American Journal of Psychiatry*. 2006;163(9):1519–1530. https://doi.org/10.1176/ajp.2006.163.9.1519.
129. Whale R, Terao T, Cowen P, Freemantle N, Geddes J. Pindolol augmentation of serotonin reuptake inhibitors for the treatment of depressive disorder: a systematic review. *Journal of Psychopharmacology*. 2010;24(4):513–520. https://doi.org/10.1177/0269881108097714.
130. Barowsky J, Schwartz TL. An evidence-based approach to augmentation and combination strategies for: treatment-resistant depression. *Psychiatry (Edgmont)*. 2006;3(7):42–61.
131. Furukawa TA, Streiner DL, Young LT. Is antidepressant–benzodiazepine combination therapy clinically more useful? *Journal of Affective Disorders*. 2001;65(2):173–177. https://doi.org/10.1016/S0165-0327(00)00254-8.
132. Dunlop BW, Davis PG. Combination treatment with benzodiazepines and SSRIs for comorbid anxiety and depression. *Primary Care Companion to the Journal of Clinical Psychiatry*. 2008;10(03):222–228. https://doi.org/10.4088/PCC.v10n0307.
133. Rizvi SJ, Sproule BA, Gallaugher L, McIntyre RS, Kennedy SH. Correlates of benzodiazepine use in major depressive disorder: the effect of anhedonia. *Journal of Affective Disorders*. 2015;187:101–105. https://doi.org/10.1016/j.jad.2015.07.040.
134. McMakin DL, Olino TM, Porta G, et al. Anhedonia predicts poorer recovery among youth with selective serotonin reuptake inhibitor treatment–resistant depression. *Journal of the American Academy of Child and Adolescent Psychiatry*. 2012;51(4):404–411. https://doi.org/10.1016/j.jaac.2012.01.011.
135. Davidson JRT. Major depressive disorder treatment guidelines in America and Europe. *The Journal of Clinical Psychiatry*. 2010;71(suppl E1):e04. https://doi.org/10.4088/JCP.9058se1c.04gry.
136. Higuchi T. Major depressive disorder treatment guidelines in Japan. *The Journal of Clinical Psychiatry*. 2010;71(suppl E1):e05. https://doi.org/10.4088/JCP.9058se1c.05gry.
137. Gelenberg AJ, Freeman MP, Markowitz JC, Rosenbaum JF, Thase ME, Trivedi MH. *Practice Guideline for the Treatment of Patients with Major Depressive Disorder*. 2010, 10.1176/appi.books.9780890423387.654001.
138. Landen M, Bjorling G, Agren H, Fahlen T. A randomized, double-blind, placebo-controlled trial of buspirone in combination with an SSRI in patients with treatment-refractory depression. *The Journal of Clinical Psychiatry*. 1998;59(12):664–668. https://doi.org/10.4088/JCP.v59n1204.
139. Appelberg BG, Syvalahti EK, Koskinen TE, Mehtonen O-P, Muhonen TT, Naukkarinen HH. Patients with severe depression may benefit from buspirone augmentation of selective serotonin reuptake inhibitors. *The Journal of Clinical Psychiatry*. 2001;62(6):448–452. https://doi.org/10.4088/JCP.v62n0608.
140. Trivedi MH, Fava M, Wisniewski SR, et al. Medication augmentation after the failure of SSRIs for depression. *The New England Journal of Medicine*. 2006;354(12):1243–1252. https://doi.org/10.1056/NEJMoa052964.

141. de Sousa RT, V.Zanetti M,R, Brunoni A, Machado-Vieira R. Challenging treatment-resistant major depressive disorder: a roadmap for improved therapeutics. *Current Neuropharmacology*. 2015;13(5):616–635. https://doi.org/10.2174/1570159x13666150630173522.
142. Malhi GS, Byrow Y, Bassett D, et al. Stimulants for depression: on the up and up? *Australian and New Zealand Journal of Psychiatry*. 2016;50(3):203–207. https://doi.org/10.1177/0004867416634208.
143. Giacobbe P, Rakita U, Lam R, Milev R, Kennedy SH, McIntyre RS. Efficacy and tolerability of lisdexamfetamine as an antidepressant augmentation strategy: a meta-analysis of randomized controlled trials. *Journal of Affective Disorders*. 2018;226:294–300. https://doi.org/10.1016/j.jad.2017.09.041.
144. Fava M. Augmentation and combination strategies in treatment-resistant depression. *The Journal of Clinical Psychiatry*. 2001;62(Suppl 18):4–11.
145. Minzenberg MJ, Carter CS. Modafinil: a review of neurochemical actions and effects on cognition. *Neuropsychopharmacology*. 2008;33(7):1477–1502. https://doi.org/10.1038/sj.npp.1301534.
146. Fava M, Thase ME, DeBattista C, Doghramji K, Arora S, Hughes RJ. Modafinil augmentation of selective serotonin reuptake inhibitor therapy in MDD partial responders with persistent fatigue and sleepiness. *Annals of Clinical Psychiatry*. 2007; 19(3):153–159. https://doi.org/10.1080/10401230701464858.
147. Vaccarino SR, McInerney SJ, Kennedy SH, Bhat V. The potential procognitive effects of modafinil in major depressive disorder. *The Journal of Clinical Psychiatry*. 2019; 80(6). https://doi.org/10.4088/JCP.19r12767.
148. Cusin C, Iovieno N, Iosifescu DV, et al. A randomized, double-blind, placebo-controlled trial of pramipexole augmentation in treatment-resistant major depressive disorder. *The Journal of Clinical Psychiatry*. 2013;74(07):e636–e641. https://doi.org/10.4088/JCP.12m08093.
149. Franco-Chaves JA, Mateus CF, Luckenbaugh DA, Martinez PE, Mallinger AG, Zarate Jr CA. Combining a dopamine agonist and selective serotonin reuptake inhibitor for the treatment of depression: a double-blind, randomized pilot study. *Journal of Affective Disorders*. 2013;149(0):319–325. https://doi.org/10.1016/j.jad.2013.02.003.
150. Raupp-Barcaro IF, Vital MA, Galduróz JC, Andreatini R. Potential antidepressant effect of amantadine: a review of preclinical studies and clinical trials. *Revista Brasileira Psiquiatria*. 2018;40(4):449–458. https://doi.org/10.1590/1516-4446-2017-2393.
151. Vale S, Espejel MA, Dominguez JC. Amantadine in depression. *Lancet*. 1971;2(7721): 437. https://doi.org/10.1016/s0140-6736(71)90153-x.
152. Rogóz Z, Skuza G, Daniel WA, Wójcikowski J, Dudek D, Wróbel A. Amantadine as an additive treatment in patients suffering from drug-resistant unipolar depression. *Pharmacological Reports*. 2007;59(6):778–784.
153. Goh KK, Chen CH, Chiu YH, Lu ML. Lamotrigine augmentation in treatment-resistant unipolar depression: a comprehensive meta-analysis of efficacy and safety. *Journal of Psychopharmacology*. 2019;33(6):700–713. https://doi.org/10.1177/0269881119844199.
154. Zarate Jr CA, Payne JL, Sporn J, et al. An open-label trial of riluzole in patients with treatment-resistant major depression. *The American Journal of Psychiatry*. 2004; 161(1):171–174. https://doi.org/10.1176/appi.ajp.161.1.171.
155. Sanacora G, Kendell SF, Levin Y, et al. Preliminary evidence of riluzole efficacy in antidepressant-treated patients with residual depressive symptoms. *Biological Psychiatry*. 2007;61(6):822–825. https://doi.org/10.1016/j.biopsych.2006.08.037.

156. Mathew SJ, Gueorguieva R, Brandt C, Fava M, Sanacora G. A randomized, double-blind, placebo-controlled, sequential parallel comparison design trial of adjunctive riluzole for treatment-resistant major depressive disorder. *Neuropsychopharmacology*. 2017;42(13):2567–2574. https://doi.org/10.1038/npp.2017.106.
157. Darji NH, Rana DA, Malhotra SD. Comparative efficacy between ketamine, memantine, riluzole and d-cycloserine in patients diagnosed with drug resistant depression: a meta-analysis. *International Journal of Basic and Clinical Pharmacology*. 2019; 8(6):1132. https://doi.org/10.18203/2319-2003.ijbcp20192174.
158. Sakurai H, Dording C, Yeung A, et al. Longer-term open-label study of adjunctive riluzole in treatment-resistant depression. *Journal of Affective Disorders*. 2019;258(May): 102–108. https://doi.org/10.1016/j.jad.2019.06.065.
159. Pharmascience Inc. *Product Monograph Pms-TOPIRAMATE*. 2018.
160. Mowla A, Kardeh E. Topiramate augmentation in patients with resistant major depressive disorder: a double-blind placebo-controlled clinical trial. *Progress in Neuro-Psychopharmacology and Biological Psychiatry*. 2011;35(4):970–973. https://doi.org/10.1016/j.pnpbp.2011.01.016.
161. Fountoulakis KN, Karavelas V, Moysidou S, et al. Efficacy of add-on pregabalin in the treatment of patients with generalized anxiety disorder and unipolar major depression with an early nonresponse to escitalopram: a double-blind placebo-controlled study. *Pharmacopsychiatry*. 2019;52(4):193–202. https://doi.org/10.1055/a-0695-9223.
162. Yasmin S, Carpenter LL, Leon Z, Siniscalchi JM, Price LH. Adjunctive gabapentin in treatment-resistant depression: a retrospective chart review. *Journal of Affective Disorders*. 2001;63(1–3):243–247. https://doi.org/10.1016/S0165-0327(00)00187-7.
163. Berlin RK, Butler PM, Perloff MD. Gabapentin therapy in psychiatric disorders. *The Primary Care Companion for CNS Disorders*. 2015;17(5). https://doi.org/10.4088/PCC.15r01821, 10.4088/PCC.15r01821.
164. Vigo DV, Baldessarini RJ. Anticonvulsants in the treatment of major depressive disorder. *Harvard Review of Psychiatry*. 2009;17(4):231–241. https://doi.org/10.1080/10673220903129814.
165. Abbas A, Kumar N, Choudhary R. The use of cyclooxygenase-2 inhibitors in depression: a narrative review of the state of evidence. *International Journal of Basic and Clinical Pharmacology*. 2019;8(10):2349. https://doi.org/10.18203/2319-2003.ijbcp20194286.
166. Hughes MM, Connor TJ, Harkin A. Stress-related immune markers in depression: implications for treatment. *International Journal of Neuropsychopharmacology*. 2016; 19(6):pyw001. https://doi.org/10.1093/ijnp/pyw001.
167. Strawbridge R, Arnone D, Danese A, Papadopoulos A, Herane Vives A, Cleare AJ. Inflammation and clinical response to treatment in depression: a meta-analysis. *European Neuropsychopharmacology*. 2015;25(10):1532–1543. https://doi.org/10.1016/j.euroneuro.2015.06.007.
168. Na KS, Lee KJ, Lee JS, Cho YS, Jung HY. Efficacy of adjunctive celecoxib treatment for patients with major depressive disorder: a meta-analysis. *Progress in Neuro-Psychopharmacology and Biological Psychiatry*. 2014;48:79–85. https://doi.org/10.1016/j.pnpbp.2013.09.006.
169. Faridhosseini F, Sadeghi R, Farid L, Pourgholami M. Celecoxib: a new augmentation strategy for depressive mood episodes. A systematic review and meta-analysis of randomized placebo-controlled trials. *Human Psychopharmacology Clinical and Experimental*. 2014;29(3):216–223. https://doi.org/10.1002/hup.2401.

170. Köhler O,E, Benros M, Nordentoft M, et al. Effect of anti-inflammatory treatment on depression, depressive symptoms, and adverse effects a systematic review and meta-analysis of randomized clinical trials. *JAMA Psychiatry.* 2014;71(12):1381–1391. https://doi.org/10.1001/jamapsychiatry.2014.1611.
171. Castillo MFR, Murata S, Schwarz M, et al. Celecoxib augmentation of escitalopram in treatment-resistant bipolar depression and the effects on Quinolinic Acid. *Neurology, Psychiatry and Brain Research.* 2019;32:22–29. https://doi.org/10.1016/j.npbr.2019.03.005.
172. Kim SS, Kong PJ, Kim BS, Sheen DH, Nam SY, Chun W. Inhibitory action of minocycline on lipopolysaccharide-induced release of nitric oxide and prostaglandin E2 in BV2 microglial cells. *Archives of Pharmacal Research.* 2004;27(3):314–318. https://doi.org/10.1007/BF02980066.
173. Homsi S, Federico F, Croci N, et al. Minocycline effects on cerebral edema: relations with inflammatory and oxidative stress markers following traumatic brain injury in mice. *Brain Research.* 2009;1291:122–132. https://doi.org/10.1016/j.brainres.2009.07.031.
174. Husain MI, Chaudhry IB, Husain N, et al. Minocycline as an adjunct for treatment-resistant depressive symptoms: a pilot randomised placebo-controlled trial. *Journal of Psychopharmacology.* 2017;31(9):1166–1175. https://doi.org/10.1177/0269881117724352.
175. Raison CL, Rutherford RE, Woolwine BJ, et al. A randomized controlled trial of the tumor necrosis factor antagonist infliximab for treatment-resistant depression: the role of baseline inflammatory biomarkers. *Archives of General Psychiatry.* 2013; 70(1):31–41. https://doi.org/10.1001/2013.jamapsychiatry.4.
176. Bavaresco DV, Uggioni MLR, Ferraz SD, et al. Efficacy of infliximab in treatment-resistant depression: a systematic review and meta-analysis. *Pharmacology Biochemistry and Behavior.* 2020;188:172838. https://doi.org/10.1016/j.pbb.2019.172838.
177. Corominas H, Alegre C, Narváez J, et al. Correlation of fatigue with other disease related and psychosocial factors in patients with rheumatoid arthritis treated with tocilizumab: ACT-AXIS study. *Medicine (Baltimore).* 2019;98(26):e15947. https://doi.org/10.1097/MD.0000000000015947.
178. Figueiredo-Braga M, Cornaby C, Cortez A, et al. Influence of biological therapeutics, cytokines, and disease activity on depression in rheumatoid arthritis. *Journal of Immunology Research.* 2018;2018. https://doi.org/10.1155/2018/5954897.
179. Tiosano S, Yavne Y, Watad A, et al. Impact of tocilizumab on anxiety and depression in patients with rheumatoid arthritis [abstract]. *Arthritis and Rheumatology.* 2019; 71(suppl 10).
180. Sun Y, Wang D, Salvadore G, et al. The effects of interleukin-6 neutralizing antibodies on symptoms of depressed mood and anhedonia in patients with rheumatoid arthritis and multicentric Castleman's disease. *Brain Behavior and Immunity.* 2017;66: 156–164. https://doi.org/10.1016/j.bbi.2017.06.014.
181. Khandaker GM, Oltean BP, Kaser M, et al. Protocol for the insight study: a randomised controlled trial of singledose tocilizumab in patients with depression and low-grade inflammation. *BMJ Open.* 2018;8(9):1–9. https://doi.org/10.1136/bmjopen-2018-025333.
182. Schneider LS, Small GW, Clary CM. Estrogen replacement therapy and antidepressant response to sertraline in older depressed women. *The American Journal of Geriatric Psychiatry.* 2001;9(4):393–399. https://doi.org/10.1097/00019442-200111000-00007.
183. Morgan ML, Cook IA, Rapkin AJ, Leuchter AF. Estrogen augmentation of antidepressants in perimenopausal depression. *The Journal of Clinical Psychiatry.* 2005;66(06): 774–780. https://doi.org/10.4088/JCP.v66n0617.

184. Rasgon NL, Altshuler LL, Fairbanks LA, et al. Estrogen replacement therapy in the treatment of major depressive disorder in perimenopausal women. *The Journal of Clinical Psychiatry.* 2002;63(Suppl. 7):45–48.
185. Dias RS, Kerr-Corrêa F, Moreno RA, et al. Efficacy of hormone therapy with and without methyltestosterone augmentation of venlafaxine in the treatment of postmenopausal depression: a double-blind controlled pilot study. *Menopause.* 2006;13(2): 202–211. https://doi.org/10.1097/01.gme.0000198491.34371.9c.
186. Howland RH. Use of endocrine hormones for treating depression. *Journal of Psychosocial Nursing and Mental Health Services.* 2010;48(12):13–16. https://doi.org/10.3928/02793695-20101105-01.
187. Seidman SN, Miyazaki M, Roose SP. Intramuscular testosterone supplementation to selective serotonin reuptake inhibitor in treatment-resistant depressed men: randomized placebo-controlled clinical trial. *Journal of Clinical Psychopharmacology.* 2005; 25(6):584–588. https://doi.org/10.1097/01.jcp.0000185424.23515.e5.
188. Orengo CA, Fullerton L, Kunik ME. Safety and efficacy of testosterone gel 1% augmentation in depressed men with partial response to antidepressant therapy. *Journal of Geriatric Psychiatry and Neurology.* 2005;18(1):20–24. https://doi.org/10.1177/0891988704271767.
189. Pope HG, Amiaz R, Brennan BP, et al. Parallel-group placebo-controlled trial of testosterone gel in men with major depressive disorder displaying an incomplete response to standard antidepressant treatment. *Journal of Clinical Psychopharmacology.* 2010; 30(2):126–134. https://doi.org/10.1097/JCP.0b013e3181d207ca.
190. Salín-Pascual RJ, Rosas M, Jimenez-Genchi A, Rivera-Meza BL, Delgado-Parra V. Antidepressant effect of transdermal nicotine patches in nonsmoking patients with major depression. *The Journal of Clinical Psychiatry.* 1996;57(9):387–389.
191. George TP, Sacco KA, Vessicchio JC, Weinberger AH, Shytle RD. Nicotinic antagonist augmentation of selective serotonin reuptake inhibitor-refractory major depressive disorder: a preliminary study. *Journal of Clinical Psychopharmacology.* 2008;28(3): 340–344. https://doi.org/10.1097/JCP.0b013e318172b49e.
192. Möller HJ, Demyttenaere K, Olausson B, et al. Two Phase III randomised double-blind studies of fixed-dose TC-5214 (dexmecamylamine) adjunct to ongoing antidepressant therapy in patients with major depressive disorder and an inadequate response to prior antidepressant therapy. *The World Journal of Biological Psychiatry.* 2015;16(7): 483–501. https://doi.org/10.3109/15622975.2014.989261.
193. Drevets WC, Zarate CA, Furey ML. Antidepressant effects of the muscarinic cholinergic receptor antagonist scopolamine: a review. *Biological Psychiatry.* 2013;73(12): 1156–1163. https://doi.org/10.1016/j.biopsych.2012.09.031.
194. Ellis JS, Zarate Jr CA, Luckenbaugh DA, Furey ML. Antidepressant treatment history as a predictor of response to scopolamine: clinical implications. *Journal of Affective Disorders.* 2014;162:39–42. https://doi.org/10.1016/j.jad.2014.03.010.
195. Jaffe RJ, Novakovic V, Peselow ED. Scopolamine as an antidepressant: a systematic review. *Clinical Neuropharmacology.* 2013;36(1):24–26. https://doi.org/10.1097/WNF.0b013e318278b703.
196. Thase M, Rush A. Treatment resistant depression. In: Bloom F, Kupfer D, eds. *Psychopharmacology - 4th Generation of Progress.* 4th ed. New York: Lippincott, Williams, and Wilkins; 1995.

197. Hache G, Coudore F, Gardier AM, Guiard BP. Monoaminergic antidepressants in the relief of pain: potential therapeutic utility of triple reuptake inhibitors (TRIs). *Pharmaceuticals*. 2011;4(2):285–342. https://doi.org/10.3390/ph4020285.
198. Karp JF, Butters MA, Begley AE, et al. Safety, tolerability, and clinical effect of low-dose buprenorphine for treatment-resistant depression in midlife and older adults. *The Journal of Clinical Psychiatry*. 2014;75(8):e785–e793. https://doi.org/10.4088/JCP.13m08725.
199. Ehrich E, Turncliff R, Du Y, et al. Evaluation of opioid modulation in major depressive disorder. *Neuropsychopharmacology*. 2015;40(6):1448–1455. https://doi.org/10.1038/npp.2014.330.
200. Fava M, Memisoglu A, Thase ME, et al. Opioid modulation with buprenorphine/samidorphan as adjunctive treatment for inadequate response to antidepressants: a randomized double-blind placebo-controlled trial. *The American Journal of Psychiatry*. 2016; 173(5):499–508. https://doi.org/10.1176/appi.ajp.2015.15070921.
201. Papakostas GI, Shelton RC, Zajecka JM, et al. L-methylfolate as adjunctive therapy for SSRI-resistant major depression: results of two randomized, double-blind, parallel-sequential trials. *The American Journal of Psychiatry*. 2012;169(12):1267–1274. https://doi.org/10.1176/appi.ajp.2012.11071114.
202. De Berardis D, Orsolini L, Serroni N, et al. A comprehensive review on the efficacy of S-Adenosyl-L-methionine in Major Depressive Disorder. *CNS and Neurological Disorders - Drug Targets*. 2016;15(1):35–44. https://doi.org/10.2174/1871527314666150821103825.
203. Deacon G, Kettle C, Hayes D, Dennis C, Tucci J. Omega 3 polyunsaturated fatty acids and the treatment of depression. *Critical Reviews in Food Science and Nutrition*. 2017; 57(1):212–223. https://doi.org/10.1080/10408398.2013.876959.
204. Nasca C, Bigio B, Lee FS, et al. Acetyl-L-carnitine deficiency in patients with major depressive disorder. *Proceedings of the National Academy of Sciences of the United States of America*. 2018;115(34):8627–8632. https://doi.org/10.1073/pnas.1801609115.
205. Veronese N, Stubbs B, Solmi M, Ajnakina O, Carvalho AF, Maggi S. Acetyl-l-Carnitine supplementation and the treatment of depressive symptoms: a systematic review and meta-analysis. *Psychosom Med*. 2018;80(2):154–159. https://doi.org/10.1097/PSY.0000000000000537.
206. Tiemeier H, van Tuijl HR, Hofman A, Kiliaan AJ, Breteler MM. Plasma fatty acid composition and depression are associated in the elderly: the Rotterdam Study. *The American Journal of Clinical Nutrition*. 2003;78(1):40–46. https://doi.org/10.1093/ajcn/78.1.40.
207. Bottiglieri T, Godfrey P, Flynn T, Carney MW, Toone BK, Reynolds EH. Cerebrospinal fluid S-adenosylmethionine in depression and dementia: effects of treatment with parenteral and oral S-adenosylmethionine. *Journal of Neurology, Neurosurgery, and Psychiatry*. 1990;53(12):1096–1098. https://doi.org/10.1136/jnnp.53.12.1096.
208. Gawryluk JW, Wang J-FF, Andreazza AC, Shao L, Young LT. Decreased levels of glutathione, the major brain antioxidant, in post-mortem prefrontal cortex from patients with psychiatric disorders. *The International Journal of Neuropsychopharmacology*. 2011; 14(1):123–130. https://doi.org/10.1017/S1461145710000805.
209. Papakostas GI, Mischoulon D, Shyu I, Alpert JE, Fava M. S-adenosyl methionine (SAMe) augmentation of serotonin reuptake inhibitors for antidepressant nonresponders with major depressive disorder: a double-blind, randomized clinical trial. *The American Journal of Psychiatry*. 2010;167(8):942–948. https://doi.org/10.1176/appi.ajp.2009.09081198.

210. Krawczyk K, Rybakowski J. Augmentation of antidepressants with unsaturated fatty acids omega-3 in drug-resistant depression. *Psychiatria Polska*. 2012;46(4):585–598.
211. Berk M, Dean OM, Cotton SM, et al. The efficacy of adjunctive N -acetylcysteine in major depressive disorder. *The Journal of Clinical Psychiatry*. 2014;75(6):628–636. https://doi.org/10.4088/JCP.13m08454.
212. Kriston L, Von Wolff A, Westphal A, Hölzel LP, Härter M. Efficacy and acceptability of acute treatments for persistent depressive disorder: a network meta-analysis. *Depression and Anxiety*. 2014;31(8):621–630. https://doi.org/10.1002/da.22236.
213. Krames ES, Hunter Peckham P, Rezai AR, Aboelsaad F. What is neuromodulation?. In: Krames ES, Peckham PH, Rezai AR, eds. *Neuromodulation*. 1st ed. vol. 1. Academic Press; 2009:3–8, 10.1016/B978-0-12-374248-3.00002-1.
214. Milev RV, Giacobbe P, Kennedy SH, et al. Canadian Network for Mood and Anxiety Treatments (CANMAT) 2016 clinical guidelines for the management of adults with major depressive disorder: section 4. Neurostimulation treatments. *The Canadian Journal of Psychiatry*. 2016;61(9):561–575. https://doi.org/10.1177/0706743716660033.
215. Baldinger P, Lotan A, Frey R, Kasper S, Lerer B, Lanzenberger R. Neurotransmitters and electroconvulsive therapy. *The Journal of ECT*. 2014;30(2):116–121. https://doi.org/10.1097/YCT.0000000000000138.
216. Lyden H, Espinoza RT, Pirnia T, et al. Electroconvulsive therapy mediates neuroplasticity of white matter microstructure in major depression. *Translation Psychiatry*. 2014;4(4):e380. https://doi.org/10.1038/tp.2014.21.
217. Bouckaert F, Sienaert P, Obbels J, et al. ECT: its brain enabling effects A review of electroconvulsive therapy–induced structural brain plasticity. *The Journal of ECT*. 2014; 30(2):143–151. https://doi.org/10.1097/YCT.0000000000000129.
218. Huang Y-Z, Edwards MJ, Rounis E, Bhatia KP, Rothwell JC. Theta burst stimulation of the human motor cortex. *Neuron*. 2005;45(2):201–206. https://doi.org/10.1016/j.neuron.2004.12.033.
219. Pagnin D, De Queiroz V, Pini S, Cassano GB. Efficacy of ECT in depression: a meta-analytic review. *The Journal of ECT*. 2004;20(1):13–20. https://doi.org/10.1097/00124509-200403000-00004.
220. Health Quality Ontario. *Repetitive Transcranial Magnetic Stimulation for Treatment-Resistant Depression: A Systematic Review and Meta-Analysis of Randomized*. vol. 16. 2016.
221. Heijnen WT, Birkenhäger TK, Wierdsma AI, van den Broek WW. Antidepressant pharmacotherapy failure and response to subsequent electroconvulsive therapy. *Journal of Clinical Psychopharmacology*. 2010;30(5):616–619. https://doi.org/10.1097/JCP.0b013e3181ee0f5f.
222. Jelovac A, Kolshus E, McLoughlin DM. Relapse following successful electroconvulsive therapy for major depression: a meta-analysis. *Neuropsychopharmacology*. 2013; 38(12):2467–2474. https://doi.org/10.1038/npp.2013.149.
223. Song GM, Tian X, Shuai T, et al. Treatment of adults with treatment-resistant depression: electroconvulsive therapy plus antidepressant or electroconvulsive therapy alone? evidence from an indirect comparison meta-analysis. *Medicine (United States)*. 2015; 94(26):1–14. https://doi.org/10.1097/MD.0000000000001052.
224. Brakemeier E-L, Merkl A, Wilbertz G, et al. Cognitive-Behavioral therapy as continuation treatment to sustain response after electroconvulsive therapy in depression: a randomized controlled trial. *Biological Psychiatry*. 2014;76(3):194–202. https://doi.org/10.1016/j.biopsych.2013.11.030.

225. Semkovska M, McLoughlin DM. Objective cognitive performance associated with electroconvulsive therapy for depression: a systematic review and meta-analysis. *Biological Psychiatry.* 2010;68(6):568–577. https://doi.org/10.1016/j.biopsych.2010.06.009.
226. Fraser LM, O'Carroll RE, Ebmeier KP. The effect of electroconvulsive therapy on autobiographical memory: a systematic review. *The Journal of ECT.* 2008;24(1):10–17. https://doi.org/10.1097/YCT.0b013e3181616c26.
227. Dunne RA, McLoughlin DM. Systematic review and meta-analysis of bifrontal electroconvulsive therapy versus bilateral and unilateral electroconvulsive therapy in depression. *The World Journal of Biological Psychiatry.* 2012;13(4):248–258. https://doi.org/10.3109/15622975.2011.615863.
228. Mutz J, Vipulananthan V, Carter B, Hurlemann R, Fu CHY, Young AH. Comparative efficacy and acceptability of non-surgical brain stimulation for the acute treatment of major depressive episodes in adults: systematic review and network meta-analysis. *BMJ.* 2019;364:l1079. https://doi.org/10.1136/bmj.l1079.
229. McClintock SM, Tirmizi O, Chansard M, Husain MM. A systematic review of the neurocognitive effects of magnetic seizure therapy. *International Review of Psychiatry.* 2011;23(5):413–423. https://doi.org/10.3109/09540261.2011.623687.
230. Kayser S, Bewernick BH, Matusch A, Hurlemann R, Soehle M, Schlaepfer TE. Magnetic seizure therapy in treatment-resistant depression: clinical, neuropsychological and metabolic effects. *Psychological Medicine.* 2015;45(5):1073–1092. https://doi.org/10.1017/S0033291714002244.
231. Fitzgerald PB, Hoy KE, Elliot D, et al. A pilot study of the comparative efficacy of 100 Hz magnetic seizure therapy and electroconvulsive therapy in persistent depression. *Depression and Anxiety.* 2018;35(5):393–401. https://doi.org/10.1002/da.22715.
232. Daskalakis ZJ, Dimitrova J, McClintock SM, et al. Magnetic seizure therapy (MST) for major depressive disorder. *Neuropsychopharmacology.* 2020;45(2):276–282. https://doi.org/10.1038/s41386-019-0515-4.
233. Hallett M. Transcranial magnetic stimulation: a primer. *Neuron.* 2007;55(2):187–199. https://doi.org/10.1016/j.neuron.2007.06.026.
234. Tik M, Hoffmann A, Sladky R, et al. Towards understanding rTMS mechanism of action: stimulation of the DLPFC causes network-specific increase in functional connectivity. *Neuroimage.* 2017;162:289–296. https://doi.org/10.1016/j.neuroimage.2017.09.022.
235. Fitzgerald PB, Daskalakis ZJ. The mechanism of action of rTMS. In: Fitzgerald PB, Daskalakis ZJ, eds. *Repetitive Transcranial Magnetic Stimulation Treatment for Depressive Disorders: A Practical Guide.* New York: Springer Berlin Heidelberg; 2013:13–27. https://doi.org/10.1007/978-3-642-36467-9.
236. Noda Y, Silverstein WK, Barr MS, et al. Neurobiological mechanisms of repetitive transcranial magnetic stimulation of the dorsolateral prefrontal cortex in depression: a systematic review. *Psychological Medicine.* 2015;45(16):3411–3432. https://doi.org/10.1017/S0033291715001609.
237. Sehatzadeh S, Daskalakis ZJ, Yap B, et al. Unilateral and bilateral repetitive transcranial magnetic stimulation for treatment-resistant depression: a meta-analysis of randomized controlled trials over 2 decades. *Journal of Psychiatry and Neuroscience.* 2019;44(3):151–163. https://doi.org/10.1503/jpn.180056.
238. Gross M, Nakamura L, Pascual-Leone A, Fregni F. Has repetitive transcranial magnetic stimulation (rTMS) treatment for depression improved? A systematic review and meta-analysis comparing the recent vs. the earlier rTMS studies. *Acta Psychiatrica Scandinavica.* 2007;116(3):165–173. https://doi.org/10.1111/j.1600-0447.2007.01049.x.

239. Lam RW, Chan P, Wilkins-Ho M, Yatham LN. Repetitive transcranial magnetic stimulation for treatment-resistant depression: a systematic review and metaanalysis. *The Canadian Journal of Psychiatry*. 2008;53(9):621–631. https://doi.org/10.1177/070674370805300909.
240. Berlim MT, Van Den Eynde F, Tovar-Perdomo S, Daskalakis ZJ. Response, remission and drop-out rates following high-frequency repetitive transcranial magnetic stimulation (rTMS) for treating major depression: a systematic review and meta-analysis of randomized, double-blind and sham-controlled trials. *Psychological Medicine*. 2014;44(2):225–239. https://doi.org/10.1017/S0033291713000512.
241. Bretlau LG, Lunde M, Lindberg L, Undén M, Dissing S, Bech P. Repetitive transcranial magnetic stimulation (rTMS) in combination with escitalopram in patients with treatment-resistant major depression. A double-blind, randomised, sham-controlled trial. *Pharmacopsychiatry*. 2008;41(2):41–47. https://doi.org/10.1055/s-2007-993210.
242. Triggs WJ, Ricciuti N, Ward HE, et al. Right and left dorsolateral pre-frontal rTMS treatment of refractory depression: a randomized, sham-controlled trial. *Psychiatry Reseach*. 2010;178(3):467–474. https://doi.org/10.1016/j.psychres.2010.05.009.
243. Mogg A, Pluck G, Eranti SV, et al. A randomized controlled trial with 4-month follow-up of adjunctive repetitive transcranial magnetic stimulation of the left prefrontal cortex for depression. *Psychological Medicine*. 2008;38(3):323–333. https://doi.org/10.1017/S0033291707001663.
244. Huang Y-Z, Rothwell JC. The effect of short-duration bursts of high-frequency, low-intensity transcranial magnetic stimulation on the human motor cortex. *Clinical Neurophysiology*. 2004;115(5):1069–1075. https://doi.org/10.1016/j.clinph.2003.12.026.
245. Berlim MT, McGirr A, Rodrigues dos Santos N, Tremblay S, Martins R. Efficacy of theta burst stimulation (TBS) for major depression: an exploratory meta-analysis of randomized and sham-controlled trials. *Journal of Psychiatric Research*. 2017;90:102–109. https://doi.org/10.1016/j.jpsychires.2017.02.015.
246. Bakker N, Shahab S, Giacobbe P, et al. RTMS of the dorsomedial prefrontal cortex for major depression: safety, tolerability, effectiveness, and outcome predictors for 10 Hz versus intermittent theta-burst stimulation. *Brain Stimulation*. 2015;8(2):208–215. https://doi.org/10.1016/j.brs.2014.11.002.
247. Blumberger DM, Vila-Rodriguez F, Thorpe KE, et al. Effectiveness of theta burst versus high-frequency repetitive transcranial magnetic stimulation in patients with depression (THREE-D): a randomised non-inferiority trial. *Lancet*. 2018;391(10131):1683–1692. https://doi.org/10.1016/S0140-6736(18)30295-2.
248. Li CT, Chen MH, Juan CH, et al. Efficacy of prefrontal theta-burst stimulation in refractory depression: a randomized sham-controlled study. *Brain*. 2014;137(7):2088–2098. https://doi.org/10.1093/brain/awu109.
249. Li CT, Chen MH, Juan CH, et al. Effects of prefrontal theta-burst stimulation on brain function in treatment-resistant depression: a randomized sham-controlled neuroimaging study. *Brain Stimulation*. 2018;11(5):1054–1062. https://doi.org/10.1016/j.brs.2018.04.014.
250. Duprat R, Desmyter S, Rudi DR, et al. Accelerated intermittent theta burst stimulation treatment in medication-resistant major depression: a fast road to remission? *Journal of Affective Disorders*. 2016;200:6–14. https://doi.org/10.1016/j.jad.2016.04.015.

251. Benabid AL, Pollak P, Hoffmann D, et al. Long-term suppression of tremor by chronic stimulation of the ventral intermediate thalamic nucleus. *Lancet*. 1991;337(8738): 403–406. https://doi.org/10.1016/0140-6736(91)91175-T.
252. Benabid AL, Pollak P, Gao D, et al. Chronic electrical stimulation of the ventralis intermedius nucleus of the thalamus as a treatment of movement disorders. *Journal of Neurosurgery*. 1996;84(2):203–214. https://doi.org/10.3171/jns.1996.84.2.0203.
253. Kumar R, Lozano AM, Kim YJ, et al. Double-blind evaluation of subthalamic nucleus deep brain stimulation in advanced Parkinson's disease. *Neurology*. 1998;51(3): 850–855. https://doi.org/10.1212/WNL.51.3.850.
254. Benabid AL. Deep brain stimulation for Parkinson's disease. *Current Opinion in Neurobiology*. 2003;13(6):696–706. https://doi.org/10.1016/j.conb.2003.11.001.
255. Mayberg HS, Lozano AM, Voon V, et al. Deep brain stimulation for treatment-resistant depression. *Neuron*. 2005;45(5):651–660. https://doi.org/10.1016/j.neuron.2005.02.014.
256. Holtzheimer PE, Hamani C. Deep brain stimulation for treatment-resistant depression. In: Vitek JL, ed. *Deep Brain Stimulation: Technology and Applications*. vol. 2. London: Future Medicine Ltd; 2014:64–75.
257. Drobisz D, Damborská A. Deep brain stimulation targets for treating depression. *Behavioural Brain Research*. 2019;359:266–273. https://doi.org/10.1016/j.bbr.2018.11.004.
258. Lozano AM, Mayberg HS, Giacobbe P, Hamani C, Craddock RC, Kennedy SH. Subcallosal cingulate gyrus deep brain stimulation for treatment-resistant depression. *Biological Psychiatry*. 2008;64(6):461–467. https://doi.org/10.1016/j.biopsych.2008.05.034.
259. Kennedy SH, Giacobbe P, Rizvi SJ, et al. Deep brain stimulation for treatment-resistant depression: follow-up after 3 to 6 years. *The American Journal of Psychiatry*. 2011; 168(5):502–510. https://doi.org/10.1176/appi.ajp.2010.10081187.
260. Puigdemont D, Pérez-Egea R, Portella MJ, et al. Deep brain stimulation of the subcallosal cingulate gyrus: further evidence in treatment-resistant major depression. *International Journal of Neuropsychopharmacology*. 2012;15(1):121–133. https://doi.org/10.1017/S1461145711001088.
261. Holtzheimer PE, Kelley ME, Gross RE, et al. Subcallosal cingulate deep brain stimulation for treatment-resistant unipolar and bipolar depression. *Archives of General Psychiatry*. 2012;69(2):150–158. https://doi.org/10.1001/archgenpsychiatry.2011.1456.
262. Holtzheimer PE, Husain MM, Lisanby SH, et al. Subcallosal cingulate deep brain stimulation for treatment-resistant depression: a multisite, randomised, sham-controlled trial. *Lancet Psychiatry*. 2017;4(11):839–849. https://doi.org/10.1016/S2215-0366(17)30371-1.
263. Puigdemont D, Portella MJ, Pérez-Egea R, et al. A randomized double-blind crossover trial of deep brain stimulation of the subcallosal cingulate gyrus in patients with treatment-resistant depression: a pilot study of relapse prevention. *Journal of Psychiatry Neuroscience*. 2015;40(4):224–231. https://doi.org/10.1503/jpn.130295.
264. Riva-Posse P, Choi KS, Holtzheimer PE, et al. A connectomic approach for subcallosal cingulate deep brain stimulation surgery: prospective targeting in treatment-resistant depression. *Molecular Psychiatry*. 2018;23(4):843–849. https://doi.org/10.1038/mp.2017.59.
265. Merkl A, Aust S, Schneider G-H, et al. Deep brain stimulation of the subcallosal cingulate gyrus in patients with treatment-resistant depression: a double-blinded randomized controlled study and long-term follow-up in eight patients. *Journal of Affective Disorders*. 2018;227:521–529. https://doi.org/10.1016/j.jad.2017.11.024.

266. Coenen VA, Bewernick BH, Kayser S, et al. Superolateral medial forebrain bundle deep brain stimulation in major depression: a gateway trial. *Neuropsychopharmacology.* 2019;44(7):1224–1232. https://doi.org/10.1038/s41386-019-0369-9.
267. Bewernick BH, Kayser S, Sturm V, Schlaepfer TE. Long-term effects of nucleus accumbens deep brain stimulation in treatment-resistant depression: evidence for sustained efficacy. *Neuropsychopharmacology.* 2012;37(9):1975–1985. https://doi.org/10.1038/npp.2012.44.
268. Jiménez F, Velasco F, Salin-Pascual R, et al. A patient with a resistant major depression disorder treated with deep brain stimulation in the inferior thalamic peduncle. *Neurosurgery.* 2005;57(3):585–593. https://doi.org/10.1227/01.NEU.0000170434.44335.19.
269. Jiménez F, Velasco F, Salín-Pascual R, et al. Neuromodulation of the inferior thalamic peduncle for major depression and obsessive compulsive disorder. *Acta Neurochirurgica Supplement.* 2007;97(Pt 2):393–398. https://doi.org/10.1007/978-3-211-33081-4_44.
270. Sartorius A, Henn FA. Deep brain stimulation of the lateral habenula in treatment resistant major depression. *Medical Hypotheses.* 2007;69(6):1305–1308. https://doi.org/10.1016/j.mehy.2007.03.021.
271. Raymaekers S, Luyten L, Bervoets C, Gabriëls L, Nuttin B. Deep brain stimulation for treatment-resistant major depressive disorder: a comparison of two targets and long-term follow-up. *Translation Psychiatry.* 2017;7(10):e1251. https://doi.org/10.1038/tp.2017.66.
272. Fitzgerald PB, Segrave R, Richardson KE, et al. A pilot study of bed nucleus of the stria terminalis deep brain stimulation in treatment-resistant depression. *Brain Stimulation.* 2018;11(4):921–928. https://doi.org/10.1016/j.brs.2018.04.013.
273. Bergfeld IO, Mantione M, Hoogendoorn MLC, et al. Deep brain stimulation of the ventral anterior limb of the internal capsule for treatment-resistant depression: a randomized clinical trial. *JAMA Psychiatry.* 2016;73(5):456–464. https://doi.org/10.1001/jamapsychiatry.2016.0152.
274. van der Wal JM, Bergfeld IO, Lok A, et al. Long-term deep brain stimulation of the ventral anterior limb of the internal capsule for treatment-resistant depression. *Journal of Neurology, Neurosurgery and Psychiatry.* 2020;91(2):189–195. https://doi.org/10.1136/jnnp-2019-321758.
275. Malone DA, Dougherty DD, Rezai AR, et al. Deep brain stimulation of the ventral capsule/ventral striatum for treatment-resistant depression. *Biological Psychiatry.* 2009;65(4):267–275. https://doi.org/10.1016/j.biopsych.2008.08.029.
276. Kilian HM, Meyer DM, Bewernick BH, Spanier S, Coenen VA, Schlaepfer TE. Discontinuation of superolateral medial forebrain bundle deep brain stimulation for treatment-resistant depression leads to critical relapse. *Biological Psychiatry.* 2019;85(6): e23–e24. https://doi.org/10.1016/j.biopsych.2018.07.025.
277. Carreno FR, Frazer A. Vagal nerve stimulation for treatment-resistant depression. *Neurotherapeutics.* 2017;14(3):716–727. https://doi.org/10.1007/s13311-017-0537-8.
278. Bottomley JM, LeReun C, Diamantopoulos A, Mitchell S, Gaynes BN. Vagus nerve stimulation (VNS) therapy in patients with treatment resistant depression: a systematic review and meta-analysis. *Comprehensive Psychiatry.* 2020;98:152156. https://doi.org/10.1016/j.comppsych.2019.152156.
279. Aaronson ST, Carpenter LL, Conway CR, et al. Vagus nerve stimulation therapy randomized to different amounts of electrical charge for treatment-resistant depression:

acute and chronic effects. *Brain Stimulation.* 2013;6(4):631–640. https://doi.org/10.1016/j.brs.2012.09.013.

280. Aaronson ST, Sears P, Ruvuna F, et al. A 5-year observational study of patients with treatment-resistant depression treated with vagus nerve stimulation or treatment as usual: comparison of response, remission, and suicidality. *The American Journal of Psychiatry.* 2017;174(7):640–648. https://doi.org/10.1176/appi.ajp.2017.16010034.
281. Schrader LM, Cook IA, Miller PR, Maremont ER, DeGiorgio CM. Trigeminal nerve stimulation in major depressive disorder: first proof of concept in an open pilot trial. *Epilepsy and Behavior.* 2011;22(3):475–478. https://doi.org/10.1016/j.yebeh.2011.06.026.
282. Elger G, Hoppe C, Falkai P, Rush AJ, Elger CE. Vagus nerve stimulation is associated with mood improvements in epilepsy patients. *Epilepsy Research.* 2000;42(2–3):203–210. https://doi.org/10.1016/S0920-1211(00)00181-9.
283. Cook IA, Schrader LM, DeGiorgio CM, Miller PR, Maremont ER, Leuchter AF. Trigeminal nerve stimulation in major depressive disorder: acute outcomes in an open pilot study. *Epilepsy and Behavior.* 2013;28(2):221–226. https://doi.org/10.1016/j.yebeh.2013.05.008.
284. Shiozawa P, Duailibi MS, da Silva ME, Cordeiro Q. Trigeminal nerve stimulation (TNS) protocol for treating major depression: an open-label proof-of-concept trial. *Epilepsy and Behavior.* 2014;39:6–9. https://doi.org/10.1016/j.yebeh.2014.07.021.
285. Shiozawa P, da Silva ME, Netto GTM, Taiar I, Cordeiro Q. Effect of a 10-day trigeminal nerve stimulation (TNS) protocol for treating major depressive disorder: a phase II, sham-controlled, randomized clinical trial. *Epilepsy and Behavior.* 2015;44:23–26. https://doi.org/10.1016/j.yebeh.2014.12.024.
286. Stagg CJ, Nitsche MA. Physiological basis of transcranial direct current stimulation. *Neuroscience.* 2011;17(1):37–53. https://doi.org/10.1177/1073858410386614.
287. Meron D, Hedger N, Garner M, Baldwin DS. Transcranial direct current stimulation (tDCS) in the treatment of depression: systematic review and meta-analysis of efficacy and tolerability. *Neuroscience and Biobehavioral Reviews.* 2015;57:46–62. https://doi.org/10.1016/j.neubiorev.2015.07.012.
288. Palm U, Schiller C, Fintescu Z, et al. Transcranial direct current stimulation in treatment resistant depression: a randomized double-blind, placebo-controlled study. *Brain Stimulation.* 2012;5(3):242–251. https://doi.org/10.1016/j.brs.2011.08.005.
289. Blumberger DM, Tran LC, Fitzgerald PB, Hoy KE, Daskalakis ZJ. A randomized double-blind sham-controlled study of transcranial direct current stimulation for treatment-resistant major depression. *Frontiers in Psychiatry.* 2012;3. https://doi.org/10.3389/fpsyt.2012.00074. Article 74.
290. Bennabi D, Nicolier M, Monnin J, et al. Pilot study of feasibility of the effect of treatment with tDCS in patients suffering from treatment-resistant depression treated with escitalopram. *Clinical Neurophysiology.* 2015;126(6):1185–1189. https://doi.org/10.1016/j.clinph.2014.09.026.
291. Li M-S, Du X-D, Chu H-C, et al. Delayed effect of bifrontal transcranial direct current stimulation in patients with treatment-resistant depression: a pilot study. *BMC Psychiatry.* 2019;19(1):180. https://doi.org/10.1186/s12888-019-2119-2.
292. Van Bronswijk S, Moopen N, Beijers L, Ruhe HG, Peeters F. Effectiveness of psychotherapy for treatment-resistant depression: a meta-analysis and meta-regression. *Psychological Medicine.* 2019;49(3):366–379. https://doi.org/10.1017/S003329171800199X.

293. Li J-M, Zhang Y, Su W-J, et al. Cognitive behavioral therapy for treatment-resistant depression: a systematic review and meta-analysis. *Psychiatry Research*. 2018;268: 243–250. https://doi.org/10.1016/j.psychres.2018.07.020.
294. Muttoni S, Ardissino M, John C. Classical psychedelics for the treatment of depression and anxiety: a systematic review. *Journal of Affective Disorders*. 2019;258:11–24. https://doi.org/10.1016/j.jad.2019.07.076.
295. Kurland AA. LSD in the supportive care of the terminally ill cancer patient. *Journal of Psychoactive Drugs*. 1985;17(4):279–290. https://doi.org/10.1080/02791072.1985.10524332.
296. Baumeister D, Barnes G, Giaroli G, Tracy D. Classical hallucinogens as antidepressants? A review of pharmacodynamics and putative clinical roles. *Therapeutic Advances in Psychopharmacology*. 2014;4(4):156–169. https://doi.org/10.1177/2045125314527985.
297. Carhart-Harris RL, Bolstridge M, Day CMJ, et al. Psilocybin with psychological support for treatment-resistant depression: six-month follow-up. *Psychopharmacology*. 2018;235(2):399–408. https://doi.org/10.1007/s00213-017-4771-x.
298. Carhart-Harris RL, Bolstridge M, Rucker J, et al. Psilocybin with psychological support for treatment-resistant depression: an open-label feasibility study. *Lancet Psychiatry*. 2016;3(7):619–627. https://doi.org/10.1016/S2215-0366(16)30065-7.
299. Grof S. *LSD Psychotherapy*. 4th ed. Multidisciplinary Association for Psychedelic Studies; 2008.
300. Rucker JJH, Jelen LA, Flynn S, Frowde KD, Young AH. Psychedelics in the treatment of unipolar mood disorders: a systematic review. *Journal of Psychopharmacology*. 2016;30(12):1220–1229. https://doi.org/10.1177/0269881116679368.
301. Mithoefer MC, Feduccia AA, Jerome L, et al. MDMA-assisted psychotherapy for treatment of PTSD: study design and rationale for phase 3 trials based on pooled analysis of six phase 2 randomized controlled trials. *Psychopharmacology*. 2019;236(9): 2735–2745. https://doi.org/10.1007/s00213-019-05249-5.
302. dos Santos RG, Osório FL, Crippa JAS, Riba J, Zuardi AW, Hallak JEC. Antidepressive, anxiolytic, and antiaddictive effects of ayahuasca, psilocybin and lysergic acid diethylamide (LSD): a systematic review of clinical trials published in the last 25 years. *Therapeutic Advances in Psychopharmacology*. 2016;6(3):193–213. https://doi.org/10.1177/2045125316638008.
303. Palhano-Fontes F, Andrade KC, Tofoli LF, et al. The psychedelic state induced by ayahuasca modulates the activity and connectivity of the default mode network. *PLoS One*. 2015;10(2):e0118143. https://doi.org/10.1371/journal.pone.0118143.
304. Sanches RF, De Lima Osório F, Santos RGD, et al. Antidepressant effects of a single dose of ayahuasca in patients with recurrent depression a SPECT study. *Journal of Clinical Psychopharmacology*. 2016;36(1):77–81. https://doi.org/10.1097/JCP.0000000000000436.
305. Palhano-Fontes F, Barreto D, Onias H, et al. Rapid antidepressant effects of the psychedelic ayahuasca in treatment-resistant depression: a randomized placebo-controlled trial. *Psychological Medicine*. 2019;49(4):655–663. https://doi.org/10.1017/S0033291718001356.
306. Vidal S, Gex-Fabry M, Bancila V, et al. Efficacy and safety of a rapid intravenous injection of ketamine 0.5 mg/kg in treatment-resistant major depression. *Journal of Clinical Psychopharmacology*. 2018;38(6):590–597. https://doi.org/10.1097/JCP.0000000000000960.
307. Andrade C. Ketamine for depression, 4: in what dose, at what rate, by what route, for how long, and at what frequency? *The Journal of Clinical Psychiatry*. 2017;78(7): e852–e857. https://doi.org/10.4088/JCP.17f11738.

308. Murrough JW, Iosifescu DV, Chang LC, et al. Antidepressant efficacy of ketamine in treatment-resistant major depression: a two-site randomized controlled trial. *The American Journal of Psychiatry*. 2013;170(10):1134–1142. https://doi.org/10.1176/appi.ajp.2013.13030392.
309. Zarate Jr CA, Singh JB, Carlson PJ, et al. A randomized trial of an N-methyl-D-aspartate antagonist in treatment-resistant major depression. *Archives of General Psychiatry*. 2006;63(8):856. https://doi.org/10.1001/archpsyc.63.8.856.
310. Mathew SJ, Murrough JW, aan het Rot M, Collins KA, Reich DL, Charney DS. Riluzole for relapse prevention following intravenous ketamine in treatment-resistant depression: a pilot randomized, placebo-controlled continuation trial. *International Journal of Neuropsychopharmacology*. 2010;13(1):71–82. https://doi.org/10.1017/S1461145709000169.
311. Ibrahim L, DiazGranados N, Franco-Chaves J, et al. Course of improvement in depressive symptoms to a single intravenous infusion of ketamine vs add-on riluzole: results from a 4-week, double-blind, placebo-controlled study. *Neuropsychopharmacology*. 2012;37(6):1526–1533. https://doi.org/10.1038/npp.2011.338.
312. Shiroma PR, Johns B, Kuskowski M, et al. Augmentation of response and remission to serial intravenous subanesthetic ketamine in treatment resistant depression. *Journal of Affective Disorders*. 2014;155(1):123–129. https://doi.org/10.1016/j.jad.2013.10.036.
313. aan het Rot M, Collins KA, Murrough JW, et al. Safety and efficacy of repeated-dose intravenous ketamine for treatment-resistant depression. *Biological Psychiatry*. 2010;67(2):139–145. https://doi.org/10.1016/j.biopsych.2009.08.038.
314. Murrough JW, Perez AM, Pillemer S, et al. Rapid and longer-term antidepressant effects of repeated ketamine infusions in treatment-resistant major depression. *Biological Psychiatry*. 2013;74(4):250–256. https://doi.org/10.1016/j.biopsych.2012.06.022.
315. Phillips JL, Norris S, Talbot J, et al. Single, repeated, and maintenance ketamine infusions for treatment-resistant depression: a randomized controlled trial. *The American Journal of Psychiatry*. 2019;176(5):401–409. https://doi.org/10.1176/appi.ajp.2018.18070834.
316. Diamond PR, Farmery AD, Atkinson S, et al. Ketamine infusions for treatment resistant depression: a series of 28 patients treated weekly or twice weekly in an ECT clinic. *Journal of Psychopharmacology*. 2014;28(6):536–544. https://doi.org/10.1177/0269881114527361.
317. Cusin C, Ionescu DF, Pavone KJ, et al. Ketamine augmentation for outpatients with treatment-resistant depression: preliminary evidence for two-step intravenous dose escalation. *Australian and New Zealand Journal of Psychiatry*. 2017;51(1):55–64. https://doi.org/10.1177/0004867416631828.
318. Vande Voort JL, Morgan RJ, Kung S, et al. Continuation phase intravenous ketamine in adults with treatment-resistant depression. *Journal of Affective Disorders*. 2016;206:300–304. https://doi.org/10.1016/j.jad.2016.09.008.
319. Murrough JW, Soleimani L, DeWilde KE, et al. Ketamine for rapid reduction of suicidal ideation: a randomized controlled trial. *Psychological Medicine*. 2015;45(16):3571–3580. https://doi.org/10.1017/S0033291715001506.
320. Gálvez V, Li A, Huggins C, et al. Repeated intranasal ketamine for treatment-resistant depression – the way to go? Results from a pilot randomised controlled trial. *Journal of Psychopharmacology*. 2018;32(4):397–407. https://doi.org/10.1177/0269881118760660.
321. Loo CK, Gálvez V, O'Keefe E, et al. Placebo-controlled pilot trial testing dose titration and intravenous, intramuscular and subcutaneous routes for ketamine in depression. *Acta Psychiatrica Scandinavica*. 2016;134(1):48–56. https://doi.org/10.1111/acps.12572.

322. George D, Gálvez V, Martin D, et al. Pilot randomized controlled trial of titrated subcutaneous ketamine in older patients with treatment-resistant depression. *The American Journal of Geriatric Psychiatry*. 2017;25(11):1199–1209. https://doi.org/10.1016/j.jagp.2017.06.007.
323. Bozymski KM, Crouse EL, Titus-Lay EN, Ott CA, Nofziger JL, Kirkwood CK. Esketamine: a novel option for treatment-resistant depression. Ann Pharmacother. *December*. 2019. https://doi.org/10.1177/1060028019892644, 106002801989264.
324. Zheng W, Zhou Y-L, Liu W-J, et al. Investigation of medical effect of multiple ketamine infusions on patients with major depressive disorder. *Journal of Psychopharmacology*. 2019;33(4):494–501. https://doi.org/10.1177/0269881119827811.
325. Fedgchin M, Trivedi M, Daly EJ, et al. Efficacy and safety of fixed-dose esketamine nasal spray combined with a new oral antidepressant in treatment-resistant depression: results of a randomized, double-blind, active-controlled study (TRANSFORM-1). *International Journal of Neuropsychopharmacology*. 2019;22(10):616–630. https://doi.org/10.1093/ijnp/pyz039.
326. Turner EH. Esketamine for treatment-resistant depression: seven concerns about efficacy and FDA approval. *The Lancet Psychiatry*. 2019;6(12):977–979. https://doi.org/10.1016/S2215-0366(19)30394-3.
327. Mahase E. Esketamine for treatment resistant depression is not recommended by NICE. *BMJ*. 2020;368:m329. https://doi.org/10.1136/bmj.m329.

CHAPTER

Suicide in psychiatric disorders: rates, risk factors, and therapeutics

4

Leonardo Tondo, MD, MSc [1,2,3], Ross J. Baldessarini, MD [2,3]

[1]*Director, Psychiatry, Mood Disorder Lucio Bini Center, Cagliari, Italy;* [2]*Research Associate, Psychiatry, McLean Hospital, Belmont, MA, United States;* [3]*Professor, (Neuroscience), Harvard Medical School, McLean Hospital, Belmont, MA, United States*

Introduction

Suicidal risks in the general population

Suicidal ideation, planning, attempts, and fatalities are prevalent among psychiatric patients and are of major, global clinical and public health concern.[1–3] The international suicide rate in recent years has averaged 10.5 per 100,000 person-exposure years (PEY, or per 100,000/year) [4] and may be decreasing somewhat overall.[2] However, suicide rates have been rising in recent years in some world regions, notably in North America (Canada, Mexico, and the United States), across rural and urban areas and age groups, despite marked differences in rates across regions and groups based on age, sex, and ethnicity.[5–7] In the United States, suicide rates were similar to current international levels (10.5) in the late 1990s but have since increased to 14.0 per 100,000 PEY and may even be underreported.[8] Proposed reasons for this large increase include growing prevalence of substance abuse, notably including opioids with potential lethality on overdoses.[8,9]

Quantitative relationships among the several levels of suicidal risk (ideation, planning, attempt, suicide) are important but sometimes misunderstood—particularly the large numerical distance between estimates of suicidal ideation and suicide or even attempts.[10] The total proportion of the US adult population with any level of suicidal ideation or behavior has recently averaged 6.5% (Table 4.1). Estimates of the prevalence of suicidal ideation in the same general population averaged 4.5%; approximately 1.3% of adults formed specific plans for suicide and about 0.62% made an attempt, whereas only about 0.02% committed suicide.[11] Among young adults with suicidal ideation, an estimated 13.2% attempted suicide.[12] The ratio of those with suicidal ideation to attempts was approximately 7.3, whereas that of ideation to suicides was 217 and that of attempts to suicides was about 30 (Table 4.1). The ratio of attempts/suicides can serve as an "index of lethality," smaller values indicating fewer attempts per suicide, or greater lethality. This ratio in patients with mood disorder is well below 10, compared to approximately 30 in general population samples.[13,14]

Ketamine for Treatment-Resistant Depression. https://doi.org/10.1016/B978-0-12-821033-8.00004-6

Table 4.1 Suicidal risks among 209 million US adults.

Outcome	Persons (millions)	Population rate (% [95% confidence interval])
Ideation	9.5	4.55 [3.69–5.53]
Plans	2.7	1.29 [0.85–1.87]
Attempts	1.3	0.62 [0.33–1.06]
Suicides	0.044	0.021 [0.015–0.028]
All suicidality	13.5	6.48 [6.27–6.68]
	Risk ratio	
Ideation/attempts	7.34	
Attempts/suicides	30	
Ideation/suicides	217	

Data from 33 metropolitan areas of the United States, averaged for 2013–15.
Adapted from Park-Lee E, Hedden SL, Lipari RN. Suicidal Thoughts and Behavior in 33 Metropolitan Statistical Areas Update: 2013 to 2015. CBHSQ Report. Rockville, MD: US Substance Abuse and Mental Health Services Administration; 2018. Accessible at: https://www.samhsa.gov/data/sites/default/files/report_3452/ShortReport-3452.html; accessed 22 Jan 2020.

Risk factors

Suicidal risks in psychiatric disorders

Psychiatric disorders vary widely in suicidal risks, although major mood disorders carry especially high rates.[3,15–20] Bipolar disorder (BD), as well as major depressive disorder (MDD) severe enough to require hospitalization, and substance abuse disorders bear the highest reported rates among psychiatric disorders for attempts and suicides.[3,15,18,19,21–25] The standardized mortality ratio in these high-risk clinical conditions can reach 20 times that in the age- and sex-matched general population.[13,15,26,27] Among patients with BD, suicide risk has remained high, despite the growing variety of treatments with putative mood-stabilizing effects.[22,23,28] This sustained risk surely reflects the great difficulty of effectively treating bipolar depressive and mixed manic-depressive states.[29–33] Notably, approximately one-third of patients with BD attempt suicide at least once—often in early years of illness even before diagnosis and treatment are established.[13,24,25,27,34]

In our recent study of 6050 psychiatric patients evaluated at an European psychiatric research center, we found the highest suicide rate among patients diagnosed with *Diagnostic and Statistical Manual of Mental Disorders* (Fifth Edition) *DSM-5* BD with psychotic features, type I BD, and BD with prominent mixed (manic-depressive) features. Their suicide rates were as high as 202/100,000 PEY, or nearly 20 times above international general population rates.[3] Risks were somewhat lower in type II BD, and much lower among patients with MDD, who have a wide range of morbidity over time (Table 4.2). Risks were also high with substance abuse disorders and far lower with anxiety-related disorders (Table 4.2).[3]

Table 4.2 Risks of suicidal risks in adult subjects with major mood disorders.

Disorder	Attempters (A) (per 100 PEY)	Suicides (S) (per 100 PEY)	A/S ratio
Bipolar psychotic	**1.39** [1.11–1.71]	**0.202** [0.097–0.366]	6.88 [6.06–7.86]
Bipolar-I	**0.961** [0.728–1.23]	**0.137** [0.056–0.279]	7.01 [6.01–8.26]
Bipolar mixed	**1.45** [1.09–1.86]	0.097 [0.020–0.278]	14.9 [12.3–18.3]
Bipolar-II	**0.890** [0.731–1.06]	**0.075** [0.033–0.147]	11.9 [9.57–15.0]
All bipolar	**1.08** [0.966–1.21]	**0.117** [0.078–0.169]	9.23 [7.79–11.1]
Major depression	**0.426** [0.355–0.507]	**0.050** [0.028–0.084]	8.52 [6.59–11.3]

Data are averages of individually computed rates (persons with at least one attempt or suicide per 100 patient-exposure years [PEY]) with 95% confidence interval, ranked by descending order of suicide rates, among 6050 psychiatric patients at risk for an average of 13.5 years. Rates of suicide were also high with substance abuse (0.52/100 PEY) and schizophrenia-like psychotic disorders (0.15/100 PEY) but were very low with attention, anxiety, and obsessive-compulsive disorders (≤0.01/100 PEY).
Adapted from Baldessarini RJ, Tondo L. Suicidal risks in 12 DSM-5 psychiatric disorders. J Aff Disord. 2020;271:66–73.

Risk factors for suicidal behavior

In addition to psychiatric diagnoses, suicide has been associated with various predictive or "risk factors".[19] Completed suicide and relatively violent attempts are more prevalent among men in most cultures, and in the United States, suicide rates are higher among Native and Caucasian than Hispanic or African American populations.[8,35,36] Among persons diagnosed with a major mood disorder, suicide *attempts* have been associated, notably, with previous attempts, female sex, younger age, social isolation, separations, being unmarried, hopelessness, anger or hostility, impulsive-aggressive traits, co-occurring abuse of drugs or alcohol, certain affective temperamental traits (including dysthymia and cyclothymia but not hyperthymia), and poor personal or clinical support and recent psychiatric hospitalization.[18,19,25,35–45] Suicidal risk also is associated with primary or co-occurring personality disorders, especially of the cluster B type (mainly antisocial and borderline) and cluster C type (avoidant, dependent, and obsessive-compulsive).[15,16,46–48]

Current morbid state is an important risk factor for suicides and attempts—most often depression, especially with mixed (hypomanic-dysphoric) features or agitation and with melancholic features in both BD and MDD.[18,19,31–33,38,49–52] Moreover, suicidal risk rises with time spent in depression.[53,54] In general, it is our impression that depressed mood is strongly associated with suicidal ideation, whereas mixed states and agitation are particularly associated with suicidal behavior.

Several studies compared rates of suicide attempts during various illness phases in patients with BD and those with MDD. Findings include an association with depressive initial episodes, with multiple or treatment-resistant depressions, hopelessness, and particularly high risks in depressive phases with mixed manic-depressive features in both disorders.[3,18,19,33,34,40,49,51,54–58] Co-occurring disorders associated with suicidal risk in patients with mood disorder include anxiety and eating disorders, alcohol and drug abuse, personality disorders, and insomnia.[26,47,59] Family history of suicidal behavior as well as childhood trauma, emotional abuse, and neglect also represent important predictors of later suicidal behavior.[26,37,57,60–66] A previous suicide attempt has been especially consistently identified as a major risk factor for future attempts or suicides, possibly independent of psychiatric diagnosis.[25,36,49,67–69]

Some studies have found risk for suicide attempts to be greater among patients diagnosed with BD than with unipolar MDD.[3,13,70–73] However, MDD is a highly heterogeneous syndrome and its suicide risks rise with illness severity, the presence of melancholic or mixed features, and the need for hospitalization,[13,19,33,52,74,75] as well perhaps with relatively more education.[76] Age can also affect suicidal risks in patients with mood disorder: rates of attempts were reportedly highest in patients with BD aged 20–24 years and in those with MDD at ages 35–39 years [73].

Recent comparisons of rates and predictors of lifetime risk of suicide attempt between patients with BD-I and those with BD-II found higher rates in type I cases and associations in both diagnoses of suicide attempts with female sex, co-occurring psychiatric and substance use disorders, binge eating or bulimic behavior, a lifetime history of rapid cycling (especially among men), as well as relatively young age at illness onset and indicators of an adverse illness course.[3,69,77] However, two meta-analyses found no difference in rates of suicide attempts between patients with type I BD and those with type II BD.[25,68] Our recent analysis of data pertaining to 3284 patients with mood disorder supported many of the preceding factors associated with risk of suicide or attempts, including those selectively associated with suicidal risk in patients with BD versus those with MDD. Notably prominent risk factors included diagnosis of BD, treatment with a mood stabilizer, hospitalization, substance abuse, separation or divorce, family history of suicide, mixed features, and proportion of time ill (Table 4.3).

The importance of current depression as a risk factor for suicide in both patients with MDD and those with BD deserves special emphasis. Suicidal risk can be reduced either by preventing depressive morbidity or by treating acute depressive episodes associated with suicidal ideation or behavior. This proposal is even more relevant in that depressive or dysphoric morbidity accounts for three-quarters of the 40%–50% of time ill among treated patients with BD, and virtually all the similar proportions of unsolved morbidity in patients with MDD, even when receiving clinically supervised treatments long-term.[78–80] Again, depressive states most associated with suicide have been characterized as agitated or dysphoric in both BD and MDD.[14,29,81]

Despite the substantial research on risk factors for suicide just summarized, there are surprisingly few direct comparisons of factors across subjects with different mood disorders evaluated under the same conditions, and some findings have been inconsistent.[19,34] Clarifying risk factors for individual types of patients with

Table 4.3 Risk factors for suicide or attempt in major mood disorders.

Factor	Rank by risk
With all major mood disorders	
Bipolar diagnosis	1
Treated with a mood stabilizer	2
Ever psychiatrically hospitalized	3
Substance abuse	4
Separated or divorced	5
Family history of suicide	6
Mixed (manic-depressive) features	7
Family history of bipolar disorder	8
Percent time ill	9
Mood switching	10
Unemployed	11
Age ≤25 years	12
More with bipolar disorder	
Percent time depressed	1
Total percent of time ill	2
Alcohol abuse	3
Multiple depressions (≥4)	4
Dysthymic or cyclothymic temperament	5
Early abuse or trauma	6
More with major depressive disorder	
Mood-stabilizer treatment	1
Ever psychiatrically hospitalized	2
Lack of antidepressant treatment	3
Family history of suicide	4
Antipsychotic treatment	5
Substance abuse	6
Never married	7
Early onset age (≤25 years)	8
Unemployed	9
Mixed (manic-depressive) features	10

Data based on analysis of 3284 subjects with mood disorder evaluated at an European psychiatric research center, showing most prevalent risk factors overall and preferentially for either subjects with bipolar disorder or those with major depressive disorder, ranked by strength of association. All differential associations with suicidal behavior (attempt or suicide in 371 subjects) were highly statistically significant (all $P < .01$*).*

Data adapted from Baldessarini RJ, Tondo L, Pinna M, Nuñez N, Vázquez GH. Suicidal risk factors in major affective disorders. Br J Psychiatr. 2019; 215(7):621–626.

mood disorder should enhance earlier identification of suicidal risk, support preventive interventions, and improve the treatment and prognosis of patients at risk.

Therapeutics

Challenges for studies of suicide prevention

Efforts to produce scientifically sound information concerning the effectiveness of a range of efforts to reduce suicide risk, including effects of various types of treatments used to treat psychiatric disorders, have been very limited—largely owing to the ethical challenges of designing control conditions and to the low incidence of suicidal behavior, as well as difficulties in recruiting and retaining subjects in controlled studies pertaining to suicide (Table 4.4).[82,83] Aside from reliance on nonrandomized, uncontrolled, observational studies, some information has been obtained in incidental findings concerning suicidal behavior arising from clinical studies designed for other purposes, including long-term randomized and even placebo-controlled treatment trials.[84–86] Notably, such controlled trials contribute

Table 4.4 Challenges for the design of studies to test for suicide prevention.

- Ethics usually requires comparisons of plausible alternative treatments without placebo
- Suicidal behaviors call for large samples and long exposure times to increase chance encountering rare outcomes
- Exposure time: often limited or not defined specifically for subgroups or individuals
- Treatment trials with suicide or attempt as outcome measures are rarely feasible
- Suicidal events in controlled treatment trials are typically incidental ("adverse events")
- Designing an ethical control condition while avoiding placebo is difficult
- Comparing two active treatments limits chances of observing differences
- Treatment options need to be matched for contact time and clinical support, which can influence suicidal risk
- Difficult to recruit and retain suicidal subjects in controlled trials
- Suicidal risk factors are many and complex and influence trial outcomes
- Nonspecific factors, including supportive effects of follow-up visits and contact time, can influence suicidal risks
- Trials try to but may not exclude potentially suicidal subjects to limit suicidal outcomes
- Rarity of suicidal acts encourages reliance on surrogate outcome measures
- Suicidal ideation as a surrogate outcome measure is remote from suicidal acts
- Meta-analysis can pool data across studies to increase statistical power
- Ecological and correlational studies can only suggest hypotheses for further study
- Limited market size and commercial value limit interest in treatment trials for suicide
- Alternative trial designs to address such challenges are urgently needed

Adapted and updated from Tondo L, Baldessarini RJ. Suicidal behavior in mood disorders: response to pharmacological treatment. Curr Psychiatry Rep. 2016;18(9):88–98.

to research findings that support reduction of risk of suicidal behavior in both BD and MDD during long-term treatment with lithium.[82,85–87] Efforts are being made to include standardized assessments of suicidal risks in trials of new drugs that act on the central nervous system, which may yield more reliable observations of suicide-related events.[88]

The study of events as rare as suicidal acts requires either very large samples of participants or prolonged periods, even years, of exposure and observation to facilitate encountering sufficient numbers of events to support statistical testing for differences between treatment conditions. Many studies rely on such outcomes as counts of persons with suicidal behavior over prolonged times at risk ("lifetime" risk), often without adjusting for, or sometimes without even defining, explicit exposure times for subgroups or individuals. The importance of exposure time is illustrated by potential effects of dropping out of a trial arm, which is often earlier during placebo treatment and can artifactually make an active treatment seem "riskier" than placebo because patients given an active treatment are likely to be at risk for longer times. Moreover, most randomized controlled trials are of short duration and attempt to exclude suicidal subjects, theoretically making it even more difficult to encounter suicidal events; however, the success of such efforts at exclusion has been challenged.[89] A popular alternative that can increase numbers of observations is the use of meta-analysis to pool data from many relevant studies, when none alone may provide sufficient statistical power to test for the effectiveness of an experimental intervention.

Even with adequate sampling, it is very challenging to design studies in which suicidal behavior is the explicit outcome measure, and this has rarely been done. A notable example is the InterSePT trial comparing schizophrenic patients at relatively high risk of suicide randomized to treatment with either clozapine or olanzapine—designed when potential antisuicidal effects of neither agent had been adequately tested.[90] In this trial, there were significantly fewer suicide-related events during treatment with clozapine than with olanzapine. Outcomes considered included surrogate measures such as compelling suicide threats or the need to intervene clinically to avoid apparently impending suicidal behavior, rather than actual suicides or attempts. Suicides were rare, but not fewer with clozapine treatment. Nevertheless, clozapine was considered superior overall in preventing suicide-related outcomes, and the study's findings were considered adequate to support an FDA-approved indication to prevent suicide in schizophrenia.

Other studies, including several involving ketamine, have relied on ratings of suicidal ideation as a surrogate outcome measure for suicidal behavior. However, as noted earlier (Table 4.1), suicidal ideation is only remotely associated with suicide (ideation/suicide ratio = 217) but more closely with attempts (ideation/attempt ratio = 7.34). Moreover, suicidal ideation is a self-reported and subjective measure of uncertain reliability that can range from weariness of life (passive ideation) to explicit self-destructive plans (active ideation). That is, reliance on suicidal ideation may not be an adequate surrogate measure for assessing effects of treatments on actual suicidal behavior, let alone suicide. Nevertheless, ideation is a first step toward possible suicide acts even if of short duration as in impulsive acts and is appropriately considered in clinical and research assessments of suicidal risk, particularly in patients diagnosed with a major mood or substance abuse disorder and especially when compelling, persistent, and associated with specific planning of suicidal behavior. The severity or riskiness of suicidal ideation can be ranked as follows: (1) passive wish to be dead, (2) active ideation with a method in mind but without intent to die or a suicidal plan, (3) active ideation with a method considered and probable intent to die but without a specific suicidal plan, and (4) active ideation with a method, intent to die, and a suicidal plan.[88] Even suicide attempt as an outcome measure involves considerable complexity, based on variable intent to die and on lethality or violence of means, as well as many other risk-modifying factors. Definitions and prevalence of nonfatal suicide-related behaviors and ideation, and their quantitative predictive association with suicide itself, are matters of intense discussion centered on the nomenclature distinguishing attempts, plans, and ideation, including estimates of fatal intent and the lethality of methods employed. Moreover, in the research on suicide ideation and behavior, crucial assessment of the intent to die is often neglected.[88,91,92] Noteworthy challenges for designing studies to test for effectiveness of interventions aimed at reducing suicidal risk are summarized in Table 4.4.

In general, experimental therapeutics research on suicide prevention is difficult conceptually, ethically, clinically, and quantitatively. It follows that even widely employed, seemingly plausible methods of clinically managing suicidal persons are rarely adequately supported by empirical research evidence. This circumstance leaves tension between the obligation to intervene clinically, often rapidly, despite a lack of clear empirical evidence about how best to do it.[24,82] Modern psychiatric treatments, including methods as diverse as psychotherapy, rapid hospitalization, and electroconvulsive therapy (ECT), may be useful as short-term interventions but lack evidence of reducing *long-term* suicide risk.[1,24,93–95] It is also notable that only modest proportions of persons committing suicide were receiving *any* clinical care at the time of their deaths.[8,96] As is considered next, evidence pertaining to potential antisuicidal effects of treatment with particular classes of interventions, let alone specific treatments, has been limited, inconsistent, and largely inconclusive.

Antidepressants

The strong association of depressive and dysphoric morbidity with suicide in psychiatric patients suggests that treatment with drugs with demonstrated antidepressant effects may reduce suicidal risk.[28] Evidence of both short- and long-term clinical benefits of antidepressant treatment in nonbipolar MDD is substantial, although its effects on suicidal risk remain uncertain.[27,28,97–101] In contrast, the efficacy and safety of antidepressant treatment for bipolar depression remains controversial and the treatment is inadequately studied for short-term and especially for long-term use, may have emotionally destabilizing effects long-term, and lacks explicit regulatory approval for use in BD.[102–104] There also may be increased suicidal risk with antidepressants in some cases nominally considered to represent depression, but especially when dysphoric agitation, anger, restlessness, irritability, insomnia, or behavioral disinhibition are present, particularly with co-occurring substance abuse and in young patients.[1,24,33,103,105–111] Such forms of depression may be considered "mixed states," or depression with "mixed features," based on broadened *DSM-5* criteria.[93] Studies of antidepressant treatment of various designs have yielded inconsistent evidence concerning suicides or attempts, particularly because suicidal behavior rarely was an explicit, predefined outcome measure. Instead, evidence of lower suicidal risk during treatment with an antidepressant versus placebo is based on the questionable use of suicide-related items in depression rating scales, which are weighted toward suicidal ideation rather than behavior and are likely to be confounded by overall clinical impressions of clinical status.[103,110,112–117]

Lower rates of suicide with greater apparent clinical use of antidepressants identified by prescription rates have been found in some correlational, pharmacoepidemiological, or ecological studies, notably in some Nordic countries and the United States, but not in several neighboring countries or broadly internationally.[27,99,116,118] This beneficial effect has been attributed to the increased use of second-generation antidepressants but in the United Sates and Sweden similarly declining suicide rates were found at least a decade before the introduction of fluoxetine as the first clinically successful modern, relatively safe, antidepressant in the late 1980s and the subsequent sharply increased clinical use of second-generation antidepressants.[27,116]

Studies involving largely retrospective observations of large cohorts of patients with depression, sometimes with case-control comparisons, have yielded inconsistent and inconclusive findings.[27,119,120] In such studies, suicidal behavior usually was noted incidentally as an adverse event and not as an explicit, planned outcome measure. Moreover, findings from such studies can be confounded by the preferential use of antidepressants for more severe depression, presumably representing greater suicidal risk (confounding by indication). In one clinical follow-up study,[14] we found an overall rate of suicidal ideation or acts of 16% at intake, with a 2.2 times greater risk in depressed patients with BD than in those with MDD. Based on regular clinical assessments, 81% of patient-subjects considered suicidal at intake became nonsuicidal during treatment and follow-up and only 0.5% of initially nonsuicidal subjects reported new suicidal thoughts, with no new attempts or suicides.

Randomized controlled trials (RCTs) should provide the best information about effects of antidepressant treatment on suicidal risks, but individual trials are limited in number of participants and in exposure times, in the face of relatively rare events. Moreover, most have been based on incidental and passively acquired, nonexplicit assessments of suicidal outcomes, following efforts to exclude potentially suicidal subjects (Table 4.4). Nevertheless, rates of suicidal behaviors may be at least as high in controlled trials as in cohort studies of patients with depressive disorder—perhaps not surprising in that trials involve acutely depressed persons.[89,121] Suicide rates pooled across several recent large meta-analyses of RCTs of modern and older antidepressants versus placebo were similar with all treatments and averaged 862/100,000 PEY,[97,114,122] or more than 80 times above the recent average international general population rate of 10.5/100,000 PEY and nearly 20 times above an estimated rate of 50/100,000 in clinically managed outpatients diagnosed with a depressive disorder.[13] Another caveat is that the high observed rates of suicidal acts from meta-analyses of controlled trials may exaggerate risks by *annualizing* observed rates based on brief exposure times (typically 6−12 weeks) in most trials for acute depression. Some meta-analyses have found only minor differences in rates of suicidal behaviors between depressed patients treated with antidepressants versus placebo, and others have detected somewhat greater risks with antidepressants than with placebo controls. Inexplicitly, based on retrospective analyses, findings have included increased risks in juveniles and young adults, but decreased risks in older adults, and no difference overall.[110,115,123,124] These analyses assume that the trials analyzed remained well randomized despite the risk of different dropout rates between trial arms and that temporal exposures in both drug and placebo arms remained well balanced throughout the trials, but such critical details are rarely reported. They also assume that surrogate measures of suicidal ideation or even minor self-injurious behaviors, as "adverse events," are fairly and comparably ascertained in different treatment groups and that they have important predictive value for suicide itself. As discussed earlier, all these assumptions are questionable.

To recapitulate, research on the potential effects of antidepressants on suicidal risks presents important and difficult methodological problems. It is fair to state that currently available research findings derived from observations of hundreds of thousands of subjects treated with antidepressants do not provide sufficiently rigorous and consistent information to support firmly either an increase or a decrease of suicide ideation or behavior in patients with mood disorder. Still open is the possibility that increased suicidal ideation and possibly also suicide attempts may be increased with antidepressant treatment in juvenile and young-adult patients, but decreased in older adults; however, this retrospective impression has not been supported with prospective trials using explicit suicide assessments.

Lithium

Suicidal risk was reduced during long-term treatment of patients with BD with lithium in several [67,125−127] but not all studies.[128,129] Supporting this association are meta-analyses and reviews, as well as several randomized, placebo-controlled efficacy trials not specifically designed to test for effects on suicide risk.[24,67,85,98,125,130−134] A rare RCT found a substantial but statistically nonsignificant difference in rates of suicidal acts over 12 months among patients randomized to lithium versus placebo, in which all three suicides followed randomization to placebo treatment.[135] In meta-analyses of data from 34 trials, we considered suicidal behavior in patients treated long-term with lithium for a major mood disorder, involving more than 110,000 person-years of risk. Risks of suicide and attempts were five times lower during treatment with lithium among patients with recurrent mood disorders overall, and even more (sixfold) among those diagnosed with BD.[125,127] We estimated the number needed to treat (NNT) at 23 (95% confidence interval [CI], 21–25) patients treated with lithium to avoid one life-threatening or fatal suicidal act—a relatively large NNT, which probably reflects the low frequency of suicidal acts. We also found that rates of suicidal acts increased by 20-fold within several months after discontinuing lithium maintenance treatment and were twice greater with abrupt or rapid versus gradual discontinuation (over $\geq$2 weeks), later returning to levels encountered before lithium treatment had started.[136] In addition, in eight studies of patients diagnosed with recurrent unipolar MDD (involving a total of 2434 patient-years), the risk of suicide or attempts was four times lower with lithium versus alternatives that included anticonvulsants.[87] Based on these and other findings, an authoritative European Psychiatric Association review recommended the use of long-term lithium treatment to reduce the risk of suicidal behavior in patients with BD.[137]

With the exception of clozapine for schizophrenia,[90] no other treatment has regulatory approval of an indication for an antisuicidal effect, including lithium. A limitation of support for lithium, even based on RCTs, is that available studies rely on incidental findings from trials designed to test for clinical efficacy but not explicitly for suicidal risks. An additional potential limitation of all studies of therapeutic effects is that patients who accept, tolerate, and sustain particular treatments long-term may be self-selected and may not entirely represent all clinically encountered patients. On the other hand, only patients who accept and tolerate treatment for at least several months can be considered for the analysis of its effectiveness.

A common feature of patients who appear to benefit from long-term treatment with lithium or clozapine is that they require and receive especially close clinical monitoring owing to medical risks associated with these treatments. This extra attention may provide added support and facilitate early identification of emerging symptoms that can lead to suicidal behavior. This possibility was not supported in the InterSePT trial for schizophrenia patients, in which clinician contact time was similar between treatment options.[90] However, we found that various indices of greater accessibility to clinical care were closely correlated with lower state suicide rates in the United States.[138]

The effectiveness of lithium treatment in preventing suicide is likely to be associated with reduced impulsivity and aggressiveness associated with depression or dysphoric agitated, mixed states.[130,139–142] Alternatively, lithium may have specific effects against suicide independent of its mood-stabilizing actions,[143] as suicidal risk has been reduced even among patients whose primary mood symptoms had not responded well to lithium.[137,141] Finally, the apparent, major beneficial effect of lithium treatment on the risk of suicide and attempts may be superior to any such effect of anticonvulsants proposed as mood stabilizers,[84,86] and tests and comparisons with second-generation antipsychotic drugs are needed.

Anticonvulsants

Research that directly compares suicidal risks during treatment with proved or putative mood-stabilizing anticonvulsants to lithium remains quite limited.[86,129,144] However, at least two studies found nearly threefold lower average risks of suicidal behavior with lithium than with carbamazepine or valproate among patients with BD or schizoaffective disorder.[145,146] In a meta-analysis,[84] we compared protective effects against suicidal behavior of lithium and several mood-stabilizing anticonvulsants (mainly valproate and some use of carbamazepine or lamotrigine) in six direct comparisons (half involved randomized assignments to treatments) that included more than 30,000 patients who were at risk longer with lithium than with an anticonvulsant (31 vs. 19 months). The observed rate of suicidal acts averaged 0.3% per year during treatment with lithium versus 0.9% per year with anticonvulsants, to yield a meta-analytically pooled risk ratio of 2.86 (95% CI, 2.29−3.57; $P < .0001$) or nearly threefold superiority favoring lithium over the few anticonvulsants tested in this way. Nevertheless, anticonvulsants may have some beneficial effects on suicidal behavior.[147,148]

The FDA conducted a meta-analysis of placebo-controlled trials involving 11 anticonvulsants. This analysis found *more* prevalent suicidal ideation and behavior with anticonvulsants than with placebo in patients with epilepsy but not in psychiatric patients probably associated with the different nature of the illnesses.[149] The lack of effect among psychiatric patients was further supported by other studies.[150−155] In a Danish study of over 16,600 persons sampled for 6 years,[156] addition of valproate or lithium yielded similarly lower suicidal risks than treatment only with antipsychotics, and lithium and valproate had similar associations with suicidal behavior in other studies.[129,150,157] These various findings indicate that research on anticonvulsants and suicidal risk remains inconsistent and inconclusive.

Antipsychotics

Most studies of associations between treatment with antipsychotic medicines and suicidal risk involve patients with schizophrenia or schizoaffective disorder. First-generation neuroleptic drugs are far less studied for effects on suicidal behavior than modern, second-generation antipsychotics (SGAs). A study based on more than 10,000 patients with psychotic disorder found no statistical difference in relatively short-term risk of suicides and attempts during treatment with modern or older antipsychotics versus placebo.[158] However, another large study found that mortality from all causes, as well as suicide, was more prevalent among patients with psychotic disorder *not* treated with antipsychotic drugs.[159]

The first US FDA-approved treatment of any kind with an antisuicide indication was clozapine for schizophrenic patients,[160] based mainly on a large randomized trial (InterSePT) comparing clozapine with olanzapine among schizophrenic patients at high suicidal risk.[90] This remarkable trial found greater prolongation of time to interventions for emerging suicidal risk and reduced rates of suicide attempts, but *no* reduction of mortality in patients treated with clozapine was observed, with very few suicides with either treatment. A subsequent trial in schizophrenic patients also found a more beneficial effect against suicidal behavior of clozapine compared to risperidone, quetiapine, and olanzapine.[161]

An emerging approach to treat patients with mood disorder, especially BD, is to employ SGAs.[162] Some of these drugs have substantial and growing evidence for their efficacy and safety in the treatment of bipolar depression, which has been notoriously difficult to treat otherwise with antidepressants, lithium, and mood-stabilizing anticonvulsants [103,163] and is strongly associated with suicidal behavior (Table 4.2). Several SGAs have demonstrated efficacy in bipolar depression generally [164] and in broadly defined "mixed" depression with co-occurring hypomanic features, as now defined by *DSM-5* criteria, with very little risk of inducing mania.[165,166] These SGAs include cariprazine, lurasidone, olanzapine combined with fluoxetine, and quetiapine, which are used either alone or in combination with lithium or a mood-stabilizing anticonvulsant except carbamazepine which can increase their metabolic clearance. Most established antipsychotic agents also are effectively antimanic, although lurasidone remains untested. Combined efficacy for both mania and bipolar depression indicates an extra degree of clinical safety of such treatments, particularly when used in agitated-dysphoric mixed manic-depressive states with very high suicidal risks.[103,167]

Specific research evidence remains sparse as to whether SGAs are associated with reduced risk of suicidal behavior in BD or other disorders. Clozapine has some evidence of clinical effectiveness in BD, especially against mania, including for patients who have not responded satisfactorily to other treatments [168,169] and those with psychotic features.[170] However, whether antisuicidal actions of clozapine in schizophrenia also extend to BD remains untested. This unusually effective antipsychotic agent requires further evaluation for potential antisuicidal effects in patients with mood disorder.

There is suggestive evidence that a broad range of SGAs may reduce suicidal risk in schizophrenia, or at least not increase it, as well as possibly reducing all-cause mortality.[161,171–174] Evidence of reduced risk of suicidal ideation or behavior in schizophrenia patients has been associated with sertindole,[175–177] olanzapine, and risperidone.[175,178,179] However, such benefits may not be associated with long-acting, injected preparations of risperidone or paliperidone.[180] Olanzapine added to lithium or divalproex led to lower rates of suicidal ideation in patients with mixed state, type I BD than with placebo, based on an item of a depression rating scale, which may reflect overall clinical improvement and may or may not extend to suicidal behavior.[181]

In one study, discontinuing SGAs in schizophrenia patients was followed by markedly increased rates of suicide attempts.[182] Antipsychotic agents also have risks of akathisia and agitation, which can increase suicidal risks. These adverse effects have been particularly associated with some SGAs, including aripiprazole, lurasidone, ziprasidone, and even clozapine.[183,184]

In summary, treatment with antipsychotic drugs, especially clozapine, has been associated with substantial reduction of suicide-related behaviors in schizophrenia patients. In patients with mood disorder, several modern antipsychotic agents can improve both bipolar depression and mania and perhaps limit rapid cycling, with low risks of inducing agitation or mood switches, and some can also facilitate treatment of unipolar MDD. Nevertheless, such drugs require specific testing for antisuicidal effects in patients with mood disorder.

Anxiolytics and sedatives

Available evidence is limited but does not support the hypothesis that antianxiety agents alter suicidal risk in patients with anxiety disorders or other psychiatric illnesses during either short- or long-term treatment.[148] On the other hand, discontinuation of benzodiazepine treatment, especially rapidly, is a stressor and may increase suicidal risk.[185] Moreover, behavioral disinhibition associated with benzodiazepine use might increase the risk of impulsive and aggressive behaviors, particularly in combination with alcohol, and in persons with personality disorder.[185] Although it is plausible to expect beneficial effects on suicidal risk during treatment with anxiolytics, convincing research support for this hypothesis is lacking, and it is difficult to test owing to the relatively low suicidal risk associated with anxiety disorders.[3] Moreover, antianxiety agents are commonly used as secondary treatments for a variety of psychiatric disorders and are rarely investigated as a sole pharmacologic intervention.

Ketamine

As reviewed in other chapters of this book, the glutamate N-methyl-D-aspartate (NMDA) receptor antagonist ketamine administered parenterally and its active isomer, esketamine, given intranasally have growing evidence of effective and rapid reduction of depressive symptoms in depressed patients with either MDD or BD, even when other treatments have been ineffective. Such effects have been documented within minutes of administration and can persist for at least several days.[186–189] These clinical properties make ketamine a very attractive potential treatment to reduce acute suicidal risk, whereas such a long-term effect seems unlikely. Indeed, more than a dozen controlled trials have been reported that include ratings of suicidal ideation as well as changes in depressive symptoms in depressed patients, including those with MDD or BD (Table 4.5).[189]

In these trials, the dose of ketamine (usually administered intravenously [IV]) averaged 400–500 μg/kg (CI, 8–42). Control treatments were saline injections or given intranasally with an added bitter compound to maintain blinding in 53% of trials and a potent sedative agent in the other 47%. Sedative controls include midazolam at doses averaging 25 μg/kg IV, or in one trial, propofol at 1.5 mg/kg or fentanyl 2.0 ng/kg IV. Findings from these RCTs included significant reductions of suicidal ideation ratings within 12 h that were highly significant by 24 h and persisted up to 3 days, although separation of responses to ketamine versus controls was significant in only 4/10 trials individually (Table 4.5). These observations are

Table 4.5 Controlled trials of ketamine for suicidal ideation.

Days	Trials (n)	Subjects (n)		Ketamine dose [CI] (μg/kg IV)	Controls (n)	SMD (CI)	*P* value (z-score)	Successful trials (%)
		Ketamine	Controls					
≤0.5	10	183	149	456[a] [388–524]	Saline (6) Sedative (4)	**−0.51** (−1.00 to −0.03)	0.04 (2.08)	4/10 (40.0%)
0.5–1.0	10	192	201	586[a] [396–776]	Saline (4) Sedative (6)	**−0.63** (−0.99 to −0.26)	<0.001 (3.38)	4/10 (40.0%)
1–3	8	125	94	413 [308–517]	Saline (4) Sedative (4)	**−0.57** (−0.99 to −0.14)	0.009 (2.62)	2/8 (25.0%)
3–14	11	190	171	416 [335–496]	Saline (6) Sedative (5)	**−0.19** (−0.41 to 0.03)	0.09 (1.71)	1/11 (9.09%)
14–30	5	106	87	440 [273–607]	Saline (3) Sedative (2)	**−0.24** (−0.53 to 0.05)	0.10 (1.62)	0/5 (0.00%)

Sedative controls were midazolam [six trials; mean dose = 25 (CI, 8–42) μg/kg IV; propofol (1.5 mg/kg IV) or fentanyl (2.0 ng/kg IV) (one trial), and methohexital (one trial, 1 mg/kg IV). Successful *trials individually yielded statistically significant reduction of suicidal ideation with ketamine versus controls. No study reported on suicidal behavior (attempt or suicide); reductions of suicidal ideation were associated with reduction of depressive symptoms and persisted for less than 1 week after ketamine treatment.*

Data (means with 95% CIs) are recalculated from information reported in a comprehensive, meta-analytic review by Witt et al.[189] *that included data from 16 trials (with depressed subjects, including patients with bipolar disorder in five subjects). The estimated number needed to treat is three to four at SMD = 0.50–0.60 but >12 at SMD ≤0.25.*CI, *confidence interval;* IV, *intravenous;* SMD, *standardized mean difference in change of suicidal ideation ratings between ketamine and controls*

[a] *One trial used intranasal esketamine (dose to 84 mg vs. saline with a bitter chemical), not included in mean dose reported.*

encouraging but suicidal ideation is both very distant from suicidal *behavior* and is likely to covary with the overall clinical improvement of depression. Use of ketamine for bipolar depression requires additional studies because there have been some case reports of newly arising mania in association with ketamine treatment, especially with higher doses.[190–192]

Psychotherapy and other interventions

Other clinical interventions have been employed empirically with the aim of reducing the risk of suicide; however, the extent of their scientifically credible assessment has been variable and the design of comparison control conditions has often been difficult. They include rapid hospitalization, which is often a clinically necessary intervention for an acutely suicidal patient. Of note, the days and weeks following psychiatric hospitalization carry greatly increased risks of suicide and attempts, especially when the index hospitalization was to prevent impending suicidal behavior.[193–195] ECT is considered clinically to be highly effective and is recommended for treatment of acutely suicidal patients, but research to support long-lasting suicide-preventing effects of even repeated ECT is very limited.[196–199] Other brain-stimulating treatments, including repeated transcranial magnetic stimulation, may be effective for the treatment of major depression, but their effects on the risk of suicidal behavior remain to be tested.[200]

Psychotherapies that have been studied for their ability to reduce suicidal risks notably include cognitive behavior therapy (CBT) and dialectical behavior therapy (DBT).[201,202] In 10–12 randomized trials, CBT has been found to reduce recurrences of self-injurious behaviors for 6–12 months compared to treatment as usual, with a nonsignificant reduction of suicides during follow-up over 2 years (Table 4.6).[202] DBT has been evaluated mainly among persons diagnosed with borderline personality disorder, for which research findings are encouraging but not definitive, and they have not been extended to patients with major mood disorders.[46,48]

Conclusions

This chapter considered risks and predictive risk factors for suicide and suicide attempts, with an emphasis on major mood disorders, and considered various treatment interventions that may reduce suicidal risk. We pointed out that the association of suicidal ideation with suicide attempts is quantitatively low, and its association with suicide very remote (Table 4.1). Suicidal ideation also requires assessment of subjective phenomena that may not be detected or quantified reliably. These considerations limit the value of suicidal ideation as a seemingly simple and relatively safe surrogate measure for testing suicide risk-reducing effects. Findings reviewed so far underscored the clinical importance of differentiating depression and associated suicidal risks in BD from MDD and of identifying depression with agitation or other mixed (hypomanic) features, or melancholic characteristics—with all of which suicidal risks are higher (Tables 4.2 and 4.3).

Table 4.6 CBT versus TAU to protect against repetition of self-injurious acts and suicide.

Follow-up (months)	Trials (n)	Repeat self-injury (events/subjects) Rate [95% CI]		Risk reduction (95% CI)	Odds ratio (95% CI)	*P* value (*z*-score)
		CBT	TAU			
6	12	150/673 22.3% [19.2 −25.6]	180/644 28.0% [24.5 −31.6]	**20.4%** (15.8–25.6)	**0.54** (0.34 −0.85)	0.008 (2.64)
12	10	263/ 1101 23.9% (21.4 −26.5)	308/ 1131 27.2% (24.7 −29.9)	**13.8%** (8.50–16.6)	**0.80** (0.65 −0.98)	0.03 (2.15)
		Suicides (suicides/ subjects) Rate [95% CI]				
≤24	8	9/1168 0.77% [0.35 −1.46]	15/1185 1.27% [0.71 −2.08]	**39.4%** (30.8–48.4)	**0.66** (0.29 −1.51)	0.33 (0.98)

CBT, *cognitive behavior therapy;* OR, *odds ratio;* TAU, *treatment as usual.*

OR values are computed by random-effects meta-analysis across trials at each follow-up time. Note that the risk of repeat self-harm was reduced significantly at 6 (20.4%) and 12 months (13.8%) after CBT and that the risk of suicide was 39.4% lower but not statistically significant. In addition in three trials of dialectical behavior therapy, risk of repeated self-harm was reduced nonsignificantly at 12 months [OR = 0.36 (0.05–2.47)] (z = *1.03,* P = *.30, with limited statistical power).*

Data shown are derived and recomputed from a systematic meta-analytic review of effects of psychosocial interventions on suicidal behavior (Hawton K, Witt KG, Taylor-Salisbury TI, et al. Psychosocial interventions following self-harm in adults: systematic review and meta-analysis. Lancet Psychiatr. 2016;3(8):740–750.)

We also reviewed the enormous challenges of testing clinically plausible interventions for potential suicide risk-reducing effects, in the face of the relative rarity of suicidal behavior (Table 4.4). There are many difficulties for the design, conduct, and interpretation of studies aimed at testing for antisuicidal effects of specific treatments. The ethics of studies with suicide as a potential outcome is particularly daunting and usually precludes the use of placebo controls or comparisons with other ineffective treatments or inadequate doses of possibly effective treatments. Importantly, the infrequency of suicidal behaviors, even in high-risk clinical samples, makes it difficult statistically to reach sound conclusions from samples of modest size and with relatively brief exposure times in subjects randomly assigned to alternative treatments. Often, exposure times are neither well balanced between treatments nor specified for subgroups or individuals. In addition, suicidal risk appears to vary with age, with the duration of mood disorders, and in relation to the timing of interventions in different phases of illness. At the very least, such variations call for randomizing subjects to specific treatments and avoiding "mirror-image" comparisons of the same subjects during treatment versus without an experimental treatment. It is also possible that patients who accept, tolerate, benefit from, and continue to take a treatment for any purpose may differ in some critical ways from those who refuse or discontinue the treatment. Randomized prospective trials involving explicit outcomes based on reliable objective measures of suicidal behavior are greatly to be desired.

However, the rarity of such trials reflects the ethical, clinical, and liability challenges of efforts to test for the reduction of suicidal risks (Table 4.4). In addition, the potential commercial advantages of identifying an effective treatment are not always obvious. In particular, there is little commercial interest in lithium as an unpatentable mineral, and having an antisuicide indication for clozapine has had little effect on the already small market of this highly effective but potentially toxic agent.[160] Moreover, ethically feasible, head-to-head comparisons of similarly plausible experimental treatments aimed at preventing suicide are not likely to be favored by producers of only one of the treatments. More generally, the low frequency of suicide, itself, severely constrains market interest in treatments aimed at its prevention and challenges the prolonged maintenance of high clinical vigilance aimed at its prevention.

A major emphasis of this chapter was to consider research on therapeutics related to suicidal risk, including effects of antidepressants, lithium, anticonvulsants with mood-stabilizing effects, SGAs, ketamine, brain stimulation, and psychotherapies. It is indeed remarkable that such potential therapeutic considerations for psychotropic medicines were virtually unheard of until recent years.[203]

Whether antidepressant treatment has beneficial or adverse effects on suicidal risk, in both MDD and BD, remains unresolved despite the plausible expectation that improvement of depression should be accompanied by a reduction in suicidal risk. In part, this uncertainty may reflect the broad and clinical heterogeneous state of clinical conditions considered to represent "major depression" and the need to distinguish depression in MDD from BD, especially if presenting with mixed features. In addition, the association of suicidal risk with dysphoria, agitation, anger, aggression, and impulsivity, as well as depressed mood, strongly suggests that treatment with lithium or other mood-stabilizing medicines, or some SGAs, is likely to be safer, if not more effective than antidepressants in efforts to reduce suicidal risk, especially long-term. Antidepressants can worsen overarousal and agitation and may even increase suicide risk, at least temporarily, especially early in the treatment of young patients, including those in whom depression has not yet been identified as a manifestation of BD. In general, new use of antidepressants in patients with BD and those with MDD calls for thoughtful and responsive clinical monitoring, especially in the initial days of treatment, seeking early detection of worsening or emerging agitation, dysphoria, restlessness, psychotic symptoms, anger, and insomnia, including in mixed manic-depressive states. Such clinical circumstances call for caution in the use of antidepressants or other mood-elevating agents (including psychostimulants and corticosteroids) and consideration of mood-stabilizing or antipsychotic agents as a safer option likely to ameliorate conditions conducive to suicide (Table 4.7).

Table 4.7 Testing of treatments aimed at reducing suicidal risks.

Treatment	Timing	Findings	Limitations
Antidepressants	Short- and long-term effects on suicide risk not established.	Inconsistent findings in controlled and uncontrolled trials in MDD, little research in BD, may increase risk of nonlethal suicidality at ages <25 years but may decrease it in older adults.	Lack of actual, matched exposure times; suicidal status usually assessed passively and incidentally as an adverse effect rather than as an explicit outcome measure.
Antipsychotics	May have short-term benefits; clozapine not tested in mood disorders.	Clozapine: only FDA-recognized "antisuicidal" treatment; other second-generation antipsychotics not evaluated.	Clozapine status based on one controlled trial with no reduction in (rare) suicides; tested only in schizophrenia.
Anxiolytics and sedatives	May be beneficial short term.	Very limited inconclusive research.	Potential disinhibition with increased suicidal risk; risk of abuse/dependence.
Anticonvulsants	Short- and long-term effects not established.	Anticonvulsants may be less effective than lithium; valproate most studied.	Suicidal behaviors incidental, not explicit outcomes.
Lithium	Probably effective long term, not short term.	Consistent decrease of suicide risk in nonrandomized studies and also in placebo-controlled trials in which suicidal acts are incidental outcomes.	Incidentally identified outcomes. Risk of self-selection by acceptance and tolerance of treatment.
Ketamine	Brief reduction of suicidal ideation for ca. 1 week.	Reduced suicidal ideation versus saline or sedative controls as depression improves.	No information about suicidal behavior; effects parallel brief mood improvement.
Psychotherapy	CBT reduced self-injury recurrences up to 12 months	CBT reduced recurrences of self-injury significantly up to 12 months by 14%–20%; DBT was ineffective.	Testing of DBT lacked statistical power.
ECT	Short-term benefit.	Effective as an emergency measure.	No apparent long-term benefits.
Hospitalization	Short-term benefit.	Effective as an emergency measure.	Untested for long-term benefits.

CBT, *cognitive-behavior therapy;* DBT, *dialectical behavior therapy;* ECT, *electroconvulsive therapy.*
Both CBT and DBT were compared to treatment as usual.
Adapted and updated from Table 37.2 (p 424) in.Tondo L, Baldessarini RJ. Suicide in bipolar disorder. Chapt 37 in the Bipolar Book: History, Neurobiology, and Treatment. Yildiz A, Nemeroff C, Ruiz P, editors. New York: Oxford University Press, 2015, pp 509–528.

Research on other medicinal treatments, including antipsychotic drugs and sedative anxiolytics, mood-stabilizing agents other than lithium, and ketamine, is even less well developed than treatment with antidepressants. Of all antipsychotic agents, only clozapine has substantial support for reducing risks of suicidal behaviors, at least in schizophrenic patients. Among mood-stabilizing treatments, lithium has a large body of research supporting its ability to reduce the risk of suicide and attempts with long-term use. Several SGAs have evidence of antidepressant effects, including in BD, which encourage assessment of their potential beneficial effects on suicidal behavior. Research with ketamine is encouraging for short-term reduction of depressive symptoms, even with otherwise treatment-resistant depression, along with associated decreases of suicidal ideation, but this interesting agent lacks evidence of reducing the risk of suicide or attempts, particularly for prolonged periods.

Some forms of psychotherapy, especially CBT, have evidence of reducing risk of repeated self-injury and perhaps of suicide, but require further study. Other widely employed interventions, including ECT and rapid hospitalization, are probably effective in preventing suicide in acute crises but lack evidence of long-term benefit. Moreover, the days following discharge from psychiatric hospitalization carry greatly increased risks of suicide and attempts, particularly without adequate and secure aftercare planning and implementation.

Finally, the preceding overview underscores the conclusion that research support for specific therapeutic interventions aimed at reducing suicidal risk in patients with mood disorder remains limited. Indeed, lack of therapeutic investigation in this area stands in striking disproportion to the growing urgency of the problem of suicide. Treatments with evidence of value, including clozapine for schizophrenia or lithium for major mood disorders, seem to be most useful for long-term reduction of suicidal risk, whereas ECT and rapid hospitalization probably can be helpful in acute suicidal crises, despite lack of evidence for their long-term preventive effects against suicidal behavior. Nevertheless, the pressing need for effective clinical management of potentially suicidal patients makes it essential to rely on clinical experience, including repeated direct assessments of suicidal risk, especially in patients at increased risk, with skillful and sensitive application of direct and supportive personal interventions in as protective an environment as possible.

Acknowledgments

Supported in part by the Aretæus Association (to LT), a grant from the Bruce J. Anderson Foundation and the McLean Private Donors Research Fund (to RJB).

References

1. Simon RI, Hales RE. *Textbook of Suicide Assessment and Management*. 2nd ed. Washington, DC: American Psychiatric Press; 2012.

2. Turecki G, Brent DA, Gunnell D, et al. Suicide and suicidal risk: a primer. *Nat Rev Dis Prim*. 2019;5(1):74–95.
3. Baldessarini RJ, Tondo L. Suicidal risks in 12 DSM-5 psychiatric disorders. *J Aff Disord*. 2020;271:66–73.
4. WHO (World Health Organization). *International Suicide Rates*; 2020. Accessible at https://www.who.int/gho/mental_health/suicide_rates/en/. Accessed January 22, 2020.
5. Hedegaard H, Curtin SC, Warner M. Suicide mortality in the United States, 1999–2017. *NCHS Data Brief*. 2018;330(11):1–8.
6. Plemmons G, Hall M, Doupnik S, et al. Hospitalization for suicide ideation or attempt: 2008–2015. *Pediatrics*. 2018;141(6):e20172426–e20172435.
7. Rossen LM, Hedegaard H, Khan D, Warner M. County-level trends in suicide rates in the US, 2005–2015. *Am J Prev Med*. 2018;55(1):72–79.
8. CDC (US Centers for Disease Control and Prevention). *National Suicide Rates*; 2020. Accessible at https://www.cdc.gov/nchs/products/databriefs/db330.htm. Accessed January 22, 2020.
9. Bohnert ASP, Ilgen MA. Understanding links among opioid use, overdose, and suicide. *N Engl J Med*. 2019;380(1):71–79.
10. Kessler RC, Berglund P, Borges G, Nock M, Wang PS. Trends in suicide ideation, plans, gestures, and attempts in the United States, 1990–1992 to 2001–2003. *J Am Med Assoc*. 2005;293(20):2487–2495.
11. Park-Lee E, Hedden SL, Lipari RN. *Suicidal Thoughts and Behavior in 33 Metropolitan Statistical Areas Update: 2013 to 2015. CBHSQ Report*. Rockville, MD: US Substance Abuse and Mental Health Services Administration; 2018. Accessible at: https://www.samhsa.gov/data/sites/default/files/report_3452/ShortReport-3452.html. Accessed January 22, 2020.
12. Han B, Compton WM, Blanco C, Colpe L, Huang L, McKeon R. National trends in the prevalence of suicidal ideation and behavior among young adults and receipt of mental health care among suicidal young adults. *J Am Acad Child Adolesc Psychiatr*. 2018; 57(1):20–27.
13. Tondo L, Baldessarini RJ. Suicidal risks among 2826 major affective disorder patients. *Acta Psychiatr Scand*. 2007;116(6):419–428.
14. Tondo L, Lepri B, Baldessarini RJ. Suicidal status during antidepressant treatment in 789 Sardinian patients with major affective disorder. *Acta Psychiatr Scand*. 2008; 118(2):106–115.
15. Harris EC, Barraclough B. Suicide as an outcome for mental disorders: meta-analysis. *Br J Psychiatr*. 1997;170(3):205–228.
16. Harris EC, Barraclough B. Excess mortality of mental disorder. *Br J Psychiatr*. 1998; 173(3):11–53.
17. Rihmer Z, Kiss K. Bipolar disorders and suicidal behavior. *Bipolar Disord*. 2002; 4(Suppl 1):21–25.
18. Hawton K, Sutton L, Haw C, Sinclair J, Harriss L. Suicide and attempted suicide in bipolar disorder: systematic review of risk factors. *J Clin Psychiatr*. 2005;66(6):693–704.
19. Baldessarini RJ, Tondo L, Pinna M, Nuñez N, Vázquez GH. Suicidal risk factors in major affective disorders. *Br J Psychiatr*. 2019;215(7):621–626.
20. Baldessarini RJ. Epidemiology of suicide: recent developments. *Epidemiol Psychiatr Sci*. 2020;29(11):e71–e73.
21. Goodwin FK, Jamison KR. *Manic-Depressive Illness: Bipolar Disorders and Recurrent Depression*. 2nd ed. New York: Oxford University Press; 2007.

22. Schaffer A, Isometsä ET, Tondo L, et al. International Society for Bipolar Disorders Task Force on Suicide: meta-analyses and meta-regression of correlates of suicide attempts and suicide deaths in bipolar disorder. *Bipolar Disord*. 2015a;17(1):1–16.
23. Schaffer A, Isometsä ET, Tondo L, et al. Epidemiology, neurobiology and pharmacological interventions related to suicide deaths and suicide attempts in bipolar disorder: Part I of a report of the International Society for Bipolar Disorders Task Force on Suicide in Bipolar Disorder. *Austr N Z J Psychiatr*. 2015b;49(9):785–802.
24. Tondo L, Baldessarini RJ. Suicide in bipolar disorder. In: Yildiz A, Nemeroff C, Ruiz P, eds. *Chapt 37 in* the Bipolar Book: History, Neurobiology, and Treatment. New York: Oxford University Press; 2015:509–528.
25. Tondo L, Pompili M, Forte A, Baldessarini RJ. Suicide attempts in bipolar disorders: comprehensive review of 101 reports. *Acta Psychiatr Scand*. 2016;133(3):174–186.
26. Simon GE, Hunkeler E, Fireman B, Lee JY, Savarino J. Risk of suicide attempt and suicide death in patients treated for bipolar disorder. *Bipolar Disord*. 2007;9(5): 526–530.
27. Tondo L, Isacsson G, Baldessarini RJ. Suicidal behavior in bipolar disorder. *CNS Drugs*. 2003;17(7):491–511.
28. Baldessarini RJ. *Chemotherapy in Psychiatry*. 3rd ed. New York: Springer Press; 2013.
29. Rihmer A, Gonda X, Balazs J, Faludi G. Importance of depressive mixed states in suicidal behavior. *Neuropsychopharmacol Hung*. 2008;10(1):45–49.
30. Baldessarini RJ, Vieta E, Calabrese JR, Tohen M, Bowden CL. Bipolar depression: overview and commentary. *Harv Rev Psychiatr*. 2010b;18(3):143–157.
31. Saunders KE, Hawton K. Clinical assessment and crisis intervention for the suicidal bipolar disorder patient. *Bipolar Disord*. 2013;15(5):575–583.
32. Vázquez GH, Lolich M, Cabrera C, et al. Mixed symptoms in mood disorders: systematic review. *J Aff Disord*. 2018b;225(1):756–760.
33. Tondo L, Vázquez GH, Baldessarini RJ. Suicidal behavior associated with mixed features in major mood disorders. *Psychiatr Clin No Am*. 2020a;43(1):83–93.
34. Valtonen HM, Suominen K, Mantere O, Leppämäki S, Arvilommi P, Isometsä E. Suicidal behaviour during different phases of bipolar disorder. *J Affect Disord*. 2007; 97(1–3):101–107.
35. Isometsä ET. Suicidal behavior in mood disorders—who, when, and why? *Can J Psychiatry*. 2014;59(3):120–130.
36. Hansson C, Joas E, Pålsson E, Hawton K, Runeson B, Landén M. Risk factors for suicide in bipolar disorder: cohort study of 12,850 patients. *Acta Psychiatr Scand*. 2018;138(5):456–463.
37. Dalton EJ, Cate-Carter TD, Mundo E, Parikh SV, Kennedy JL. Suicide risk in bipolar patients: role of co-morbid substance use disorders. *Bipolar Disord*. 2003;5(1):58–61.
38. Oquendo MA, Currier D, Mann JJ. Prospective studies of suicidal behavior in major depressive and bipolar disorders: what is the evidence for predictive risk factors? *Acta Psychiatr Scand*. 2006;114(3):151–158.
39. Sublette ME, Carballo JJ, Moreno C, et al. Substance use disorders and suicide attempts in bipolar subtypes. *J Psychiatr Res*. 2009;43(3):230–238.
40. Undurraga J, Baldessarini RJ, Valenti M, Pacchiarotti I, Vieta E. Suicidal risk factors in bipolar I and II disorder patients. *J Clin Psychiatr*. 2012;73(6):778–782.
41. Dalca IM, McGirr A, Renaud J, Turecki G. Gender-specific suicide risk factors: case-control study of individuals with major depressive disorder. *J Clin Psychiatr*. 2013; 74(12):1209–1216.

42. Coryell W, Kriener A, Butcher B, et al. Risk factors for suicide in bipolar I disorder in two prospectively studied cohorts. *J Affect Disord*. 2016;190(1):1–5.
43. Verdolini N, Perugi G, Samalin L, et al. Aggressiveness in depression: a neglected symptom possibly associated with bipolarity and mixed features. *Acta Psychiatr Scand*. 2017;136(4):362–372.
44. Vázquez GH, Gonda X, Lolich M, Tondo L, Baldessarini RJ. Suicidal risk and affective temperaments evaluated with TEMPS-A. *Harv Rev Psychiatr*. 2018a;26(1):8–18.
45. Miller JN, Black DW. Bipolar disorder and suicide: review. *Curr Psychiatry Rep*. 2020; 22(2):6–15.
46. Linehan MM, Korslund KE, Harned MS, et al. Dialectical behavior therapy for high suicide risk in individuals with borderline personality disorder: randomized clinical trial and component analysis. *JAMA Psychiatry*. 2015;72(5):475–482.
47. Jylha PJ, Rosenstrom T, Mantere O, et al. Temperament, character, and suicide attempts in unipolar and bipolar mood disorders. *J Clin Psychiatr*. 2016;77(2):252–260.
48. Prada P, Perroud N, Rüfenacht E, Nicastro R. Strategies to deal with suicide and nonsuicidal self-injury in borderline personality disorder, the case of DBT. *Front Psychol*. 2018;9(12):2595–2601.
49. Galfalvy H, Oquendo MA, Carballo JJ, et al. Clinical predictors of suicidal acts after major depression in bipolar disorder: prospective study. *Bipolar Disord*. 2006;8(5 Pt 2):586–595.
50. Saunders KE, Hawton K. Suicidal behaviour in bipolar disorder: understanding the role of affective states. *Bipolar Disord*. 2015;17(1):24–26.
51. Tondo L, Vázquez GH, Pinna M, Vaccotto PA, Baldessarini RJ. Characteristics of depressive and bipolar patients with mixed features. *Acta Psychiatr Scand*. 2018; 138(3):243–252.
52. Tondo L, Vázquez GH, Baldessarini RJ. Melancholic and nonmelancholic depression compared. *J Affect Disord*. 2020;266:760–765.
53. Valtonen HM, Suominen K, Haukka J, et al. Differences in incidence of suicide attempts during phases of bipolar I and II disorders. *Bipolar Disord*. 2008;10(5):588–596.
54. Holma KM, Haukka J, Suominen K, et al. Differences in incidence of suicide attempts between bipolar I and II disorders and major depressive disorder. *Bipolar Disord*. 2014; 16(6):652–661.
55. Chaudhury SR, Grunebaum MF, Galfalvy HC, et al. Does first episode polarity predict risk for suicide attempt in bipolar disorder? *J Affect Disord*. 2007;104(1–3):245–250.
56. Ryu V, Jon D-I, Cho HS, et al. Initial depressive episodes affect the risk of suicide attempts in Korean patients with bipolar disorder. *Yonsei Med J*. 2010;51(5):641–647.
57. Bellivier F, Yon L, Luquiens A, et al. Suicidal attempts in bipolar disorder: results from an observational study (EMBLEM). *Bipolar Disord*. 2011;13(4):377–386.
58. Hawton K, Comabella CC, Haw C, Saunders K. Risk factors for suicide in individuals with depression: systematic review. *J Affect Disord*. 2013;147(1–3):17–28.
59. Sánchez-Gistau V, Colom F, Mane A, Romero S, Sugranyes G, Vieta E. Atypical depression is associated with suicide attempt in bipolar disorder. *Acta Psychiatr Scand*. 2009;120(1):30–36.
60. Leverich GS, Altshuler LL, Frye MA, et al. Factors associated with suicide attempts in 648 patients with bipolar disorder in the Stanley Foundation Bipolar Network. *J Clin Psychiat*. 2003;64(5):506–515.
61. Neves FS, Malloy-Diniz LF, Correa H. Suicidal behavior in bipolar disorder: what is the influence of psychiatric comorbidities? *J Clin Psychiatr*. 2009;70(1):13–18.

62. Altamura AC, Dell'Osso B, Berlin HA, Buoli M, Bassetti R, Mundo E. Duration of untreated illness and suicide in bipolar disorder: naturalistic study. *Eur Arch Psychiatr Clin Neurosci*. 2010;260(5):385–391.
63. de Abreu LN, Nery FG, Harkavy-Friedman JM, et al. Suicide attempts are associated with worse quality of life in patients with bipolar disorder type I. *Compr Psychiatr*. 2012;53(2):125–129.
64. Gonda X, Pompili M, Serafini G, et al. Suicidal behavior in bipolar disorder: epidemiology, characteristics and major risk factors. *J Affect Disord*. 2012;143(1–3):16–26.
65. Antypa N, Serretti A, Rujescu D. Serotonergic genes and suicide: a systematic review. *Eur Neuropsychopharmacol*. 2013;23(10):1125–1142.
66. de Mattos-Souza LD, Molina ML, da Silva RA, Jansen K. History of childhood trauma as risk factors to suicide risk in major depression. *Psychiatr Res*. 2016;246(12):612–616.
67. Angst J, Angst F, Gerber-Werder R, Gamma A. Suicide in 406 mood-disorder patients with and without long-term medication: 40–44 year follow-up. *Arch Suicide Res*. 2005;9(3):279–300.
68. Novick DM, Swartz HA, Frank E. Suicide attempts in bipolar I and bipolar II disorder: review and meta-analysis of the evidence. *Bipolar Disord*. 2010;12(1):1–9.
69. Goffin KC, Dell'Osso B, Miller S, et al. Different characteristics associated with suicide attempts among bipolar I versus bipolar II disorder patients. *J Psychiatr Res*. 2016;76(5):94–100.
70. Lester D. Suicidal behavior in bipolar and unipolar affective disorders: meta-analysis. *J Affect Disord*. 1993;27(2):117–121.
71. Bottlender R, Jäger M, Strauss A, Möller HJ. Suicidality in bipolar compared to unipolar depressed inpatients. *Eur Arch Psychiatr Clin Neurosci*. 2000;250(5):257–261.
72. Zalsman G, Braun M, Arendt M, et al. Comparison of the medical lethality of suicide attempts in bipolar and major depressive disorders. *Bipolar Disord*. 2006;8(5 Pt 2):558–565.
73. Pawlak J, Dmitrzak-Weglarz M, Skibinska M, et al. Suicide attempts and clinical risk factors in patients with bipolar and unipolar affective disorders. *Gen Hosp Psychiatr*. 2013;35(4):427–432.
74. Bradvik L, Mattisson C, Bogren M, Nettelbladt P. Long-term suicide risk of depression in the Lundby cohort 1947–1997: severity and gender. *Acta Psychiatr Scand*. 2008;117(3):185–191.
75. Nuñez NA, Comai S, Dumitrescu E, et al. Psychopathological and sociodemographic features in treatment-resistant unipolar depression versus bipolar depression: comparative study. *BMC Psychiatr*. 2018;18(1):68–78.
76. Wei S, Li H, Hou J, Chen W, Chen X, Qin X. Comparison of the characteristics of suicide attempters with major depressive disorder and those with no psychiatric diagnosis in emergency of general hospitals in China. *Ann Gen Psychiatr*. 2017;16(12):44–52.
77. Bobo WV, Na PJ, Geske JR, McElroy SL, Frye MA, Biernacka JM. Relative influence of individual risk factors for attempted suicide in patients with bipolar I vs. bipolar II disorder. *J Affect Disord*. 2018;225(1):489–494.
78. Baldessarini RJ, Salvatore P, Khalsa HM, et al. Morbidity in 303 first-episode bipolar I disorder patients. *Bipolar Disord*. 2010a;12(3):264–270.
79. De Dios C, Ezquiaga E, Garcia A, Soler B, Vieta E. Time spent with symptoms in a cohort of bipolar disorder outpatients in Spain: prospective, 18-month follow-up study. *J Affect Disord*. 2010;125(1–3):74–81.

80. Forte A, Baldessarini RJ, Tondo L, Vázquez GH, Pompili M, Girardi P. Long-term morbidity in bipolar-I, bipolar-II, and unipolar major depressive disorders. *J Affect Disord*. 2015;178(6):71–78.
81. Isometsä ET, Henriksson MM, Aro HM, Heikkinen ME, Kuoppasalmi KI, Lönnqvist JK. Suicide in major depression. *Am J Psychiatr.* 1994;151(4):530–536.
82. Tondo L, Baldessarini RJ. Suicidal behavior in mood disorders: response to pharmacological treatment. *Curr Psychiatry Rep*. 2016;18(9):88–98.
83. Sareen J, Isaak C, Katz LY, Bolton J, Enns MW, Stein MB. Promising strategies for advancement in knowledge of suicide risk factors and prevention. *Am J Prev Med.* 2014;47(3 Suppl 2):S257–S263.
84. Baldessarini RJ, Tondo L. Suicidal risks during treatment of bipolar disorder patients with lithium versus anticonvulsants. *Pharmacopsychiatry.* 2009;42(2):72–75.
85. Smith KA, Cipriani A. Lithium and suicide in mood disorders: updated meta-review of the scientific literature. *Bipolar Disord*. 2017;19(7):575–586.
86. Tondo L, Baldessarini RJ. Antisuicidal effects in mood disorders: are they unique to lithium? *Pharmacopsychiatry.* 2018;51(5):177–188.
87. Guzzetta F, Tondo L, Centorrino F, Baldessarini RJ. Lithium treatment reduces suicide risk in recurrent major depressive disorder. *J Clin Psychiatr.* 2007;68(3):380–383.
88. FDA (US Food and Drug Administration). *Guidance for Industry: Suicidal Ideation and Behavior: Prospective Assessment of Occurrence in Clinical Trials*. U.S. Department of Health and Human Services, Food and Drug Administration Center for Drug Evaluation and Research (CDER); 2012. www.fda.gov/downloads/Drugs/Guidances/UCM225130.pdf. Accessed February 1, 2020.
89. Baldessarini RJ, Lau WK, Sim J, Sum MY, Sim K. Suicidal risks in reports of long-term treatment trials for major depressive disorder. *Int J Neuropsychopharmacol*. 2017;20(3):281–284.
90. Meltzer HY, Alphs L, Green AI, et al. Clozapine treatment for suicidality in schizophrenia: international suicide prevention trial (InterSePT). *Arch Gene Psychiatr.* 2003;60(1):82–91.
91. Brenner LA, Breshears RE, Betthauser LM, et al. Implementation of a suicide nomenclature within two VA healthcare settings. *J Clin Psychol Med Settings*. 2011;18(2):116–128.
92. Crosby AE, Ortega LV, Melanson C. *Self-directed Violence Surveillance: Uniform Definitions and Recommended Data Elements*. Version 1.0. Atlanta, GA: Centers for Disease Control and Prevention National Center for Injury Prevention and Control Division of Violence Prevention; 2011.
93. APA (American Psychiatric Association). *Diagnostic and Statistical Manual of Mental Disorders, Fifth Revision (DSM-5)*. Washington, DC: American Psychiatric Publishing; 2013.
94. Goldsmith SK, Pellmar TC, Kleinman AM, Bunney Jr WE. *Reducing Suicide*. Washington, DC: Institute of Medicine of the US National Academies of Science; 2002.
95. Weinger RD. *Practice of Electroconvulsive Therapy: Recommendations for Treatment, Training, and Privileging: Task Force Report of the American Psychiatric Association*. 2nd ed. Washington, DC: American Psychiatric Press; 2002.
96. Ernst CL, Bird SA, Goldberg JF, Ghaemi SN. Prescription of psychotropic medications for patients discharged from a psychiatric emergency service. *J Clin Psychiatr.* 2010;67(5):720–726.
97. Khan A, Khan S, Kolts R, Brown WA. Suicide rates in clinical trials with SRIs, other antidepressants, and placebo: analysis of FDA reports. *Am J Psychiatr.* 2003;160(4):790–792.

98. Khan A, Khan SR, Hobus J, et al. Differential pattern of response in mood symptoms and suicide risk measures in severely ill depressed patients assigned to citalopram with placebo or citalopram combined with lithium: role of lithium levels. *J Psychiatr Res*. 2011;45(11):1489–1496.
99. Søndergård L, Kvist K, Lopez AG, Andersen PK, Kessing LV. Temporal changes in suicide rates for persons treated and not treated with antidepressants in Denmark during 1995–1999. *Acta Psychiatr Scand*. 2006;114(3):168–176.
100. Yatham LN, Kennedy SH, Parikh SV, et al. Canadian network for mood and anxiety treatments (CANMAT) and international society for bipolar disorders (ISBD) collaborative update for the management of patients with bipolar disorder. *Bipolar Disord*. 2013;15(1):1–44.
101. Sim K, Lau KL, Sim J, Sum MY, Baldessarini RJ. Prevention of relapse and recurrence in adults with major depressive disorder: systematic review and meta-analyses of controlled trials. *Intl J Neuropsychopharmacol*. 2015;19(2):1–13.
102. Ghaemi SN, Hsu DJ, Soldani F, Goodwin FK. Antidepressants in bipolar disorder: the case for caution. *Bipolar Disord*. 2003;5(6):421–433.
103. Pacchiarotti I, Bond DJ, Baldessarini RJ, et al. International Society for Bipolar Disorders (ISBD) task force report on antidepressant use in bipolar disorders. *Am J Psychiatr*. 2013;170(11):1249–1262.
104. Tondo L, Baldessarini RJ, Vázquez G, Lepri B, Visioli C. Clinical responses to antidepressants among 1036 acutely depressed patients with bipolar or unipolar major affective disorders. *Acta Psychiatr Scand*. 2013;127(5):355–364.
105. Koukopoulos A, Koukopoulos A. Agitated depression as a mixed state and the problem of melancholia. *Psychiatr Clin No Am*. 1999;22(3):547–564.
106. Akiskal HS, Benazzi F. Psychopathologic correlates of suicidal ideation in major depressive outpatients: is it all due to unrecognized (bipolar) depressive mixed states? *Psychopathology*. 2005;38(5):273–280.
107. Akiskal HS, Benazzi F, Perugi G, Rihmer Z. Agitated "unipolar" depression reconceptualized as a depressive mixed state: implications for the antidepressant-suicide controversy. *J Affect Disord*. 2005;85(3):245–258.
108. Baldessarini RJ, Faedda GL, Hennen J. Risk of mania with serotonin reuptake inhibitors vs. tricyclic antidepressants in children, adolescents and young adults. *Arch Pediatr Adolesc Med*. 2005;159(3):298–299.
109. Maj M, Pirozzi R, Magliano L, Fiorillo A, Bartoli L. Agitated "unipolar" major depression: prevalence, phenomenology, and outcome. *J Clin Psychiatr*. 2006;67(5):712–719.
110. Barbui C, Esposito E, Cipriani A. Selective serotonin reuptake inhibitors and risk of suicide: systematic review of observational studies. *CMAJ*. 2009;180(3):291–297.
111. Popovic D, Vieta E, Azorin JM, et al. Suicide attempts in major depressive episode: evidence from the BRIDGE-II-Mix study. *Bipolar Disord*. 2015;17(7):795–803.
112. Beasley Jr CM, Dornseif BE, Bosomworth JC, et al. Fluoxetine and suicide: meta-analysis of controlled trials of treatment for depression. *BMJ*. 1991;303(6804):685–692.
113. Tollefson GD, Rampey Jr AH, Beasley Jr CM, Enas GG, Potvin JH. Absence of a relationship between adverse events and suicidality during pharmacotherapy for depression. *J Clin Psychopharm*. 1994;14(3):163–169.
114. Acharya N, Rosen AS, Polzer JP, et al. Duloxetine: meta-analyses of suicidal behaviors and ideation in clinical trials for major depressive disorder. *J Clin Psychopharmacol*. 2006;26(6):587–594.

115. Hammad TA, Laughren TP, Racoosin JA. Suicide rates in short-term randomized controlled trials of newer antidepressants. *J Clin Psychopharmacol.* 2006;26(2): 203–207.
116. Baldessarini RJ, Tondo L, Strombom I, et al. Analysis of ecological studies of relationships between antidepressant utilization and suicidal risk. *Harvard Rev Psychiatr.* 2007; 15(4):133–145.
117. Zisook S, Trivedi MH, Warden D, et al. Clinical correlates of the worsening or emergence of suicidal ideation during SSRI treatment of depression. *J Affect Disord.* 2009;117(1–2):63–73.
118. Reseland S, Bray I, Gunnell D. Relationship between antidepressant sales and secular trends in suicide rates in the Nordic countries. *Br J Psychiatr.* 2006; 188(4):354–358.
119. Möller HJ. Antidepressants: controversies about their efficacy in depression, their effect on suicidality and their place in a complex psychiatric treatment approach. *World J Biol Psychiatr.* 2009;10(3):180–195.
120. Valuck RJ, Libby AM, Anderson HD, et al. Comparison of antidepressant classes and the risk and time course of suicide attempts in adults: propensity matched, retrospective cohort study. *Br J Psychiatr.* 2016;208(3):271–279.
121. Braun C, Bschor T, Franklin J, Baethge C. Suicides and suicide attempts during long-term treatment with antidepressants: meta-analysis of 29 placebo-controlled studies including 6934 patients with major depressive disorder. *Psychother Psychosom.* 2016;85(3):171–179.
122. Khan A, Warner HA, Brown WA. Symptom reduction and suicide risk in patients treated with placebo in antidepressant clinical trials: analysis of the FDA database. *Arch Gen Psychiatr.* 2000;57(4):311–317.
123. Noel C. Antidepressants and suicidality: history of the black-box warning, consequences, and current evidence. *Mental Health Clinician.* 2015;5(5):202–211.
124. Wise J. Antidepressants may double risk of suicide and aggression in children, study finds. *BMJ.* 2016;352(1):i545–i547.
125. Baldessarini RJ, Tondo L, Davis P, Pompili M, Goodwin FK, Hennen J. Decreased suicidal risk during long-term lithium treatment: meta-analysis. *Bipolar Disord.* 2006;8(5 Pt 2):625–639.
126. Müller-Oerlinghausen B, Ahrens B, Felber W. Suicide-preventive and mortality-reducing effect of lithium. In: Bauer M, Grof P, Müller-Oerlinghausen B, eds. *Lithium in Neuropsychiatry.* London: Informa Healthcare; 2006:79–192.
127. Baldessarini RJ, Tondo L. Lithium and suicidal risk. *Bipolar Disord.* 2008;10(1): 114–115.
128. Marangell LB, Dennehy EB, Wisniewski SR, et al. Case-control analyses of the impact of pharmacotherapy on prospectively observed suicide attempts and completed suicides in bipolar disorder. *J Clin Psychiatr.* 2008;69(6):916–922.
129. Oquendo MA, Galfalvy HC, Currier D, et al. Treatment of suicide attempters with bipolar disorder: randomized clinical trial comparing lithium and valproate in the prevention of suicidal behavior. *Am J Psychiatr.* 2011;168(10):1050–1056.
130. Coppen A, Farmer R. Suicide mortality in patients on lithium maintenance therapy. *J Affect Disord.* 1998;50(2–3):261–267.
131. Tondo L, Hennen J, Baldessarini RJ. Reduced suicide risk with long-term lithium treatment in major affective illness: meta-analysis. *Acta Psychiatr Scand.* 2001;104(3): 163–172.

132. Cipriani A, Hawton K, Stockton S, Geddes JR. Lithium in the prevention of suicide in mood disorders: updated systematic review and meta-analysis. *BMJ*. 2013;346(6): f3646–f3658.
133. Tondo L, Baldessarini RJ. Reduction of suicidal behavior in bipolar disorder patients during long-term treatment with lithium. In: Koslow SH, Ruiz P, Nemeroff CB, eds. *A Concise Guide to Understanding Suicide*. New York: Cambridge University Press; 2014:217–228.
134. Sarai SK, Mekala HM, Lippmann S. Lithium suicide prevention: brief review and reminder. *Innov Clin Neurosci*. 2018;15(11–12):30–32.
135. Lauterbach E, Felber W, Müller-Oerlinghausen B, et al. Adjunctive lithium treatment in the prevention of suicidal behavior in depressive disorders: randomized, placebo-controlled, 1-year trial. *Acta Psychiat Scand*. 2008;118(6):469–479.
136. Tondo L, Baldessarini RJ, Hennen J, Floris G, Silvetti F, Tohen M. Lithium treatment and risk of suicidal behavior in bipolar disorder patients. *J Clin Psychiatr*. 1998;59(8): 405–414.
137. Lewitzka U, Bauer M, Felber W, Müller-Oerlinghausen B. Antisuicidal effect of lithium: current state of research and its clinical implications for the long-term treatment of affective disorders. *Nervenärzt*. 2013;84(3):294–306.
138. Tondo L, Albert MJ, Baldessarini RJ. Suicide rates in relation to health-care access in the United States: an ecological study. *J Clin Psychiatr*. 2006;67(4):517–523.
139. Barraclough B. Suicide prevention, recurrent affective disorder and lithium. *Br J Psychiatr*. 1972;121(563):391–392.
140. Bschor T, Bauer M. Efficacy and mechanisms of action of lithium augmentation in refractory major depression. *Curr Pharmaceutical Design*. 2006;12(23): 2985–2992.
141. Müller-Oerlinghausen B, Lewitzka U. Lithium reduces pathological aggression and suicidality: mini-review. *Neuropsychobiology*. 2010;62(1):43–49.
142. Manchia M, Hajek T, O'Donovan C, et al. Genetic risk of suicidal behavior in bipolar spectrum disorder: analysis of 737 pedigrees. *Bipolar Disord*. 2013;15(5):496–506.
143. Kovacsics CE, Gottesman II , Gould TD. Lithium's antisuicidal efficacy: elucidation of neurobiological targets using endophenotype strategies. *Ann Rev Pharmacol Toxicol*. 2009;49(1):175–198.
144. Søndergård L, Lopez AG, Andersen PK, Kessing LV. Mood-stabilizing pharmacological treatment in bipolar disorders and risk of suicide. *Bipolar Disord*. 2008;10(1): 87–94.
145. Thies-Flechtner K, Müller-Oerlinghausen B, Seibert W, Walther A, Greil W. Effect of prophylactic treatment on suicide risk in patients with major affective disorders: data from a randomized prospective trial. *Pharmacopsychiatry*. 1996;29(3):103–107.
146. Goodwin FK, Fireman B, Simon GE, Hunkeler EM, Lee J, Revicki D. Suicide risk in bipolar disorder during treatment with lithium and divalproex. *J Am Med Assoc*. 2003; 290(11):1467–1473.
147. Yerevanian BI, Koek RJ, Mintz J. Bipolar pharmacotherapy and suicidal behavior. Part I: lithium, divalproex and carbamazepine. *J Affect Disord*. 2007;103(1–3):5–11.
148. Yerevanian BI, Choi YM. Impact of psychotropic drugs on suicide and suicidal behaviors. *Bipolar Disord*. 2013;15(5):594–621.
149. FDA, (US Food and Drug Administration). *Statistical Review and Evaluation Antiepileptic Drugs and Suicidality*; 2008. Accessible at www.fda.gov/2008-4372b1-01.pdf. Accessed February 1, 2020.

150. Gibbons RD, Hur K, Brown CH, Mann JJ. Relationship between antiepileptic drugs and suicide attempts in patients with bipolar disorder. *Arch Gen Psychiatr.* 2009;66(12): 1354–1360.
151. Mula M, Kanner AM, Schmitz B, Schachter S. Antiepileptic drugs and suicidality: expert consensus statement from the task force on therapeutic strategies of the ILAE commission on neuropsychobiology. *Epilepsia.* 2013;54(1):199–203.
152. Ferrer P, Ballarín E, Sabaté M, et al. Antiepileptic drugs and suicide: systematic review of adverse effect. *Neuroepidemiology.* 2014;42(2):107–120.
153. Siamouli M, Samara M, Fountoulakis KN. Is antiepileptic-induced suicidality a data-based class effect or an exaggeration? A comment on the literature. *Harv Rev Psychiatr.* 2014;22(6):379–381.
154. Fountoulakis KN, Gonda X, Baghai TC, et al. Report of the WPA section of pharmacopsychiatry on the relationship of antiepileptic drugs with suicidality in epilepsy. *Int J Psychiatry Clin Pract.* 2015;19(3):158–167.
155. Rissanen I, Jääskeläinen E, Isohanni M, Koponen H, Ansakorpi H, Miettunen J. Use of antiepileptic or benzodiazepine medication and suicidal ideation: northern Finland Birth Cohort 1966. *Epilepsy Behav.* 2015;46(5):198–204.
156. Smith EG, Søndergård L, Lopez AG, Andersen PK, Kessing LV. Association between consistent purchase of anticonvulsants or lithium and suicide risk: longitudinal cohort study from Denmark, 1995–2001. *J Affect Disord.* 2009;117(3):162–167.
157. Smith EG, Austin KL, Kim HM, et al. Suicide risk in Veterans Health Administration patients with mental health diagnoses initiating lithium or valproate: historical prospective cohort study. *BMC Psychiatr.* 2014;14(12):357–371.
158. Khan A, Khan SR, Leventhal RM, Brown WA. Symptom reduction and suicide risk among patients treated with placebo in antipsychotic clinical trials: an analysis of the FDA database. *Am J Psychiatr.* 2001;158(9):1449–1454.
159. Tiihonen J, Wahlbeck K, Lönnqvist J, et al. Effectiveness of antipsychotic treatments in a nationwide cohort of patients in community care after first hospitalization due to schizophrenia and schizoaffective disorder: observational follow-up study. *BMJ.* 2006;333(7561):224–229.
160. Hennen J, Baldessarini RJ. Reduced suicidal risk during treatment with clozapine: meta-analysis. *Schizophrenia Res.* 2005;73(2–3):139–145.
161. Haukka J, Tiihonen J, Härkänen T, Lönnqvist J. Association between medication and risk of suicide, attempted suicide and death in nationwide cohort of suicidal patients with schizophrenia. *Pharmacoepidemiol Drug Saf.* 2008;17(7):686–696.
162. Ahearn EP, Chen P, Hertzberg M, et al. Suicide attempts in veterans with bipolar disorder during treatment with lithium, divalproex, and atypical antipsychotics. *J Affect Disord.* 2013;145(1):77–82.
163. Vázquez GH, Tondo L, Undurraga J, Baldessarini RJ. Overview of antidepressant treatment in bipolar depression: critical commentary. *Intl J Neuropsychopharmacol.* 2013; 16(7):1673–1685.
164. Post RM. Treatment of bipolar depression: evolving recommendations. *Psychiatr Clin No Am.* 2016;39(1):11–33.
165. Poo SX, Agius M. Atypical antipsychotics in adult bipolar disorder: current evidence and updates in the NICE guidelines. *Psychiatr Danub.* 2014;26(Suppl 1):322–329.
166. Fornaro M, Stubbs B, De Berardis D, et al. Atypical antipsychotics in the treatment of acute bipolar depression with mixed features: systematic review and exploratory meta-analysis of placebo-controlled clinical trials. *Int J Mol Sci.* 2016;17(2):241–254.

167. Swann AC, Lafer B, Perugi G, et al. Bipolar mixed states: an international society for bipolar disorders task force report of symptom structure, course of illness, and diagnosis. *Am J Psychiatr.* 2013;170(1):31–42.
168. Ifteni P, Correll CU, Nielsen J, Burtea V, Kane JM, Manu P. Rapid clozapine titration in treatment-refractory bipolar disorder. *J Affect Disord.* 2014;166(9):168–172.
169. Li XB, Tang YL, Wang CY, de Leon J. Clozapine for treatment-resistant bipolar disorder: systematic review. *Bipolar Disord.* 2015;17(3):235–247.
170. Ciapparelli A, Dell'Osso L, Pini S, Chiavacci MC, Fenzi M, Cassano GB. Clozapine for treatment-refractory schizophrenia, schizoaffective disorder, and psychotic bipolar disorder: 24-month naturalistic study. *J Clin Psychiatr.* 2000;61(5):329–334.
171. Ulcickas-Yood M, Delorenze G, Quesenberry Jr CP, et al. Epidemiologic study of aripiprazole use and the incidence of suicide events. *Pharmacoepidemiol Drug Saf.* 2010; 19(11):1124–1130.
172. Kiviniemi M, Suvisaar J, Koivumaa-Honkanen H, Häkkinen U, Isohanni M, Hakko H. Antipsychotics and mortality in first-onset schizophrenia: prospective Finnish register study with 5-year follow-up. *Schizophr Res.* 2013;150(1):274–280.
173. Reutfors J, Bahmanyar S, Jönsson EG, et al. Medication and suicide risk in schizophrenia: nested case-control study. *Schizophr Res.* 2013;150(2–3):416–420.
174. Toffol E, Hätönen T, Tanskanen A, et al. Lithium is associated with decrease in all-cause and suicide mortality in high-risk bipolar patients: nationwide registry-based prospective cohort study. *J Affect Disord.* 2015;183(9):159–165.
175. Kerwin RW, Bolonna AA. Is clozapine antisuicidal? *Expert Rev Neurother.* 2004;4(2): 187–190.
176. Crocq MA, Naber D, Lader MH, et al. Suicide attempts in a prospective cohort of patients with schizophrenia treated with sertindole or risperidone. *Eur Neuropsychopharmacol.* 2010;20(12):829–838.
177. Thomas SH, Drici MD, Hall GC, et al. Safety of sertindole versus risperidone in schizophrenia: principal results of the sertindole cohort prospective study. *Acta Psychiatr Scand.* 2010;122(5):345–355.
178. Barak Y, Mirecki I, Knobler HY, Natan Z, Aizenberg D. Suicidality and second generation antipsychotics in schizophrenia patients: case-controlled retrospective study during a 5-year period. *Psychopharmacology.* 2004;175(2):215–219.
179. Reeves H, Batra S, May RS, Zhang R, Dahl DC, Li X. Efficacy of risperidone augmentation to antidepressants in the management of suicidality in major depressive disorder: randomized, double-blind, placebo-controlled pilot study. *J Clin Psychiatr.* 2008;69(8): 1228–1236.
180. Gentile S. Adverse effects associated with second-generation antipsychotic long-acting injection treatment: comprehensive systematic review. *Pharmacotherapy.* 2013;33(10): 1087–1106.
181. Houston JP, Ahl J, Meyers AL, Kaiser CJ, Tohen M, Baldessarini RJ. Reduced suicidal ideation in bipolar I disorder mixed-episode patients in a placebo-controlled trial of olanzapine combined with lithium or divalproex. *J Clin Psychiatr.* 2006;67(8): 1246–1252.
182. Herings RM, Erkens JA. Increased suicide attempt rate among patients interrupting use of atypical antipsychotics. *Pharmacoepidemiol Drug Saf.* 2003;12(5):423–424.
183. Seemüller F, Lewitzka U, Bauer M, et al. Relationship of akathisia with treatment emergent suicidality among patients with first-episode schizophrenia treated with haloperidol or risperidone. *Pharmacopsychiatry.* 2012a;45(7):292–296.
184. Seemüller F, Schennach R, Mayr A, et al. Akathisia and suicidal ideation in first-episode schizophrenia. *J Clin Psychopharmacol.* 2012b;32(5):694–698.

185. Gaertner I, Gilot C, Heidrich P, Gaertner HJ. Case control study on psychopharmacotherapy before suicide committed by 61 psychiatric inpatients. *Pharmacopsychiatry.* 2002;35(2):37–43.
186. Vázquez GH, Camino S, Tondo L, Baldessarini RJ. Potential novel treatments for bipolar depression: ketamine, fatty acids, anti-inflammatory agents, and probiotics. *CNS Neurol Disord - Drug Targets.* 2017;16(8):858–869.
187. Li KX, Loshak H. *Intravenous Ketamine for Adults with Treatment-Resistant Depression or Post-traumatic Stress Disorder: Review of Clinical Effectiveness, Cost-Effectiveness and Guidelines.* Ottawa (ON): Canadian Agency for Drugs and Technologies in Health; 2019. Oct 24.
188. Park LT, Falodun TP, Zarate Jr CA. Ketamine for treatment-resistant mood disorders. *APPI Focus.* 2019;17(1):8–12.
189. Witt K, Potts J, Hubers A, et al. Ketamine for suicidal ideation in adults with psychiatric disorders: systematic review and meta-analysis of treatment trials. *Austr N Z J Psychiatr.* 2020;54(1):39–45.
190. Banwari G, Desai P, Patidar P. Ketamine-induced affective switch in a patient with treatment-resistant depression. *Indian J Pharmacol.* 2015;47(4):454–455.
191. Lu YY, Lin CH, Lane HY. Mania following ketamine abuse. *Neuropsychiatr Dis Treat.* 2016;12(1):237–239.
192. Mandyam MC, Ahuja NK. Ketamine-Induced mania during treatment for complex regional pain syndrome. *Pain Med.* 2017;18(10):2040–2041.
193. Kessler RC, Warner CH, Ivany C, et al. Predicting suicides after psychiatric hospitalization in US army soldiers: the army study to assess risk and resilience in servicemembers (army STARRS). *JAMA Psychiatr.* 2015;72(1):49–57.
194. Chung DT, Ryan CJ, Hadzi-Pavlovic D, Singh SP, Stanton C, Large MM. Suicide rates after discharge from psychiatric facilities: systematic review and meta-analysis. *JAMA Psychiatr.* 2017;75(7):694–702.
195. Forte A, Buscajoni A, Fiorillo A, Pompili M, Baldessarini RJ. Suicidal risk following hospitalization: review. *Harv Rev Psychiatr.* 2019;27(4):209–216.
196. Sharma V. Effect of electroconvulsive therapy on suicide risk in patients with mood disorders. *Can J Psychiatr.* 2001;46(8):704–709.
197. Mehlum L, Dieserud G, Ø E, et al. Prevention of suicide: Part 1: psychotherapy, drug treatment and electroconvulsive treatment. *Oslo: Norwegian Knowl Cen Health Ser (NOKC), Report.* 2006;24.
198. Fink M, Kellner CH, McCall WV. The role of ECT in suicide prevention. *J ECT.* 2014; 30(1):5–9.
199. Jørgensen MB, Rozing MP, Kellner CH, Osler M. Electroconvulsive therapy, depression severity and mortality: data from the Danish National Patient Registry. *J Psychopharmacol.* 2020;34(3):273–279.
200. Sverak T, Ustohal L. Efficacy and safety of intensive transcranial magnetic stimulation. *Harv Rev Psychiatr.* 2018;26(1):19–26.
201. Calati R, Courtet P. Is psychotherapy effective for reducing suicide attempts and non-suicidal self-injury rates? Meta-analysis and meta-regression of literature data. *J Psychiatr Res.* 2016;79(8):8–20.
202. Hawton K, Witt KG, Taylor-Salisbury TI, et al. Psychosocial interventions following self-harm in adults: systematic review and meta-analysis. *Lancet Psychiatr.* 2016; 3(8):740–750.
203. Effects of medical interventions on suicidal behaviorBaldessarini RJ, Jamison KR, eds. *J Clin Psychiatr.* 1999;60(Suppl 2):1–122.

CHAPTER

Overview of ketamine for major depression: efficacy and effectiveness

5

Anees Bahji, MD [1,2], Gustavo H. Vazquez, MD, PhD, FRCPC [5], Elisa M. Brietzke, MD, PhD [4], Carlos A. Zarate, MD [3]

[1]*Resident, Department of Psychiatry, Queen's University School of Medicine, Kingston, ON, Canada;* [2]*Doctor, Psychiatry, Department of Public Health Sciences, Queen's University, Kingston, ON, Canada;* [3]*Chief, Experimental Therapeutics and Pathophysiology Branch, Section Neurobiology and Treatment of Mood Disorders, Division of Intramural Research Program, National Institute of Mental Health, Bethesda, MD, United States;* [4]*Professor, Kingston General Hospital, Providence Care Hospital, Queen's University School of Medicine, Kingston, ON, Canada;* [5]*Professor, Lead, Ketamine Clinic, Mood Disorders Outpatient Unit, Queen's University, Department of Psychiatry, Providence Care Hospital, Kingston, ON, Canada*

Historical background

Depression is the leading cause of disability in the world.[1,2] Although depressive symptoms may be reduced within several weeks following the initiation of treatment with conventional antidepressants, treatment resistance concerns in one-third of patients who fail to achieve recovery.[3] Over the past 20 years, ketamine, an antagonist of the *N*-methyl-D-aspartate receptor (NMDAR), has been described to have antidepressant properties.[3] Consequently, the clinical use of ketamine is increasing, and intranasal (*S*)-ketamine has recently been approved for treating depression by the Food and Drug Administration and may very well become a promising treatment in depressed patients with suicidal ideation.[3] In fact, there is increasing evidence that the antisuicidal properties of ketamine may be helpful in the treatment of other conditions. For example, there is controlled data, albeit small, for the utility of ketamine in the treatment of posttraumatic stress disorder,[4–8] There is also some preliminary uncontrolled evidence indicating that ketamine may be of therapeutic value in the treatment of cancer-related depression[9] and cancer-related neuropathic pain.[10–16] Collectively, the level of proof of efficacy remains low and more randomized controlled trials are needed to explore the efficacy and safety issues of ketamine in depression,[3] but preliminary evidence suggests that ketamine may have broader therapeutic effects than conventional antidepressants.

Ketamine has been around for more than a half century; however, its unique neurobiological profile is only starting to be uncovered.[17] First introduced as an invaluable anesthetic, there is an emerging role for ketamine in the management of treatment-refractory depression.[17] Primarily, ketamine functions as an antagonist

Ketamine for Treatment-Resistant Depression. https://doi.org/10.1016/B978-0-12-821033-8.00005-8

of the NMDA subtype of the glutamate receptor.[18] Ketamine is reported to have rapid antidepressant effects; however, there is limited understanding of the time-course of ketamine effects beyond a single infusion.[19] Furthermore, it is unclear if the NMDA mechanism or the enhancement of γ-aminobutyric acid (GABA) transmission explains ketamine's rapid antidepressant properties.[20] In the recent years, the neurosteroid antidepressant brexanolone (allopregnanolone) has provided some additional support for the GABA-mediated (GABAergic) deficit hypothesis of depression, which has implications on ketamine's physiologic mechanisms.[21] However, the precise mechanism for ketamine's therapeutic effects is still unclear, and the initial findings were specific to postpartum depression (rather than all forms of depression).[22–24] Although there was a positive trial for brexanolone in major depressive disorder and there are several ongoing trials, recent large controlled studies have not shown the oral form to separate from placebo at the primary trial endpoint.[25]

The antidepressant properties of ketamine: evidence from animal studies

Numerous clinical and preclinical studies suggest that dysfunction of the glutamatergic system is implicated in the pathophysiology of mood disorders, such as major depressive disorder and bipolar depression. Consequently, this may explain the ameliorating properties of ketamine—a glutamate receptor antagonist.[26] However, the discovery of ketamine's rapid antidepressant effects in individuals with major depression is one of the most significant findings in clinical psychopharmacology in the recent decades.[27]

The early clinical studies in human subjects were based on the foundation of rodent preclinical studies carried out in the 1990s, particularly rodents.[27] Although there is general agreement in the rodent literature that ketamine has rapidly acting, and generally sustained, antidepressant-like properties, there are also points of contention across studies, including the precise mechanism of action of this drug.[27] To date, available data best support the notion of ketamine action principally via NMDAR antagonism, transiently boosting glutamatergic (and possibly other) signaling in diverse brain circuits.[27] Ketamine is effective in animal models of treatment-resistant depression (TRD), thus corroborating the literature findings.[28] Low-dose ketamine can also treat depressionlike behaviors induced by chronic neuropathic pain in rats.[29] Furthermore, an acute single dose of ketamine relieves depressionlike behaviors in a rat model.[30] Given the role of cytokine-mediated inflammation in the pathophysiology of depression, an expanding body of scientific evidence indicates that ketamine has antiinflammatory properties, which may contribute to its antidepressant effects. For example, the rapid antidepressant effect of ketamine in rats has been associated with the downregulation of proinflammatory cytokines in the hippocampus.[31] Furthermore, ketamine appears to alleviate depressionlike behaviors via downregulating inflammatory cytokines induced by chronic restraint stress in mice.[32] Ketamine modulates hippocampal neurogenesis and proinflammatory cytokines but not stressor-induced neurochemical changes.[33] Repeated ketamine treatment induces sex-specific behavioral and neurochemical effects in mice, such as beneficial antidepressant-like effects in male mice, but it induced both anxietylike and depressionlike effects in their female counterparts.[34,35] Murine models have indicated that chronic psychologic stress interacts with ketamine treatment to modulate its effects in the forced swim test, which reinforces the relevance to stress-induced depression in humans.[36]

Open-label and nonrandomized studies on the antidepressant potential of ketamine

In clinical studies of individuals with major depressive disorder and bipolar depression, rapid reductions in depressive symptoms have been observed in response to subanesthetic doses of ketamine.[26] In turn, the findings from these studies have prompted the repurposing and/or development of other glutamatergic modulators, such as D-cycloserine, riluzole, dextromethorphan, and nitrous oxide, for antidepressant efficacy, both as monotherapy and as an adjunct to conventional monoaminergic antidepressants.[26]

Patients receiving hospice care and who received daily oral ketamine experienced a robust antidepressant and anxiolytic response with few adverse events, showing a response rate for depression similar to that found with intravenous ketamine.[37] The rapid antidepressant effects of ketamine seen in individuals with TRD appear to be predictive of a sustained effect.[19] A pilot randomized controlled trial of titrated subcutaneous ketamine in older patients with TRD demonstrated preliminary evidence for the efficacy and safety of ketamine in treating elderly depressed persons.[38]

In one study, adolescents with TRD who were administered six ketamine infusions over 2 weeks demonstrated an average decrease in the Children's Depression Rating Scale Revised (CDRS-R) of 42.5%, with 38% meeting criteria for clinical response and 23% showing sustained remission at 6-week follow-up.[39]

Ketamine for suicidality

In addition to antidepressant properties, ketamine can rapidly reduce suicidal thoughts within 1 day and for up to a few weeks in depressed patients with suicidal ideation—partially independent of its effects on mood.[40–42] The antisuicidal properties of ketamine have been previously identified in earlier works in the context of depression.[43,44] Repeated doses of open-label ketamine rapidly and robustly decreased suicidal ideation in pharmacologically treated outpatients with TRD.[45] Both intravenous ketamine and intranasal esketamine have shown great potential as potent and rapid antisuicidal agents.[40,42,46–49] Given the current limitations of most existing treatments for reducing suicidal ideations and plans in patients with moderate to severe major depression, this additional property of ketamine may be particularly helpful in the emergent management of patients in acute crisis.

Evidence from randomized controlled trials

In terms of biological gradients, low-dose ketamine (0.5 mg/kg) appears to be the optimal dose for the treatment of depression. At this dose, there is an increase in efficacy relative to lower doses (0.1−0.2 mg/kg) in terms of clinical response in major depression, while retaining a similar efficacy to higher doses (1.0 mg/kg), which are less well tolerated.[50,51]

The emergence of a novel ketamine formulation, i.e., intranasal esketamine, has diversified the existing options for delivery of the rapid antidepressant effects of ketamine in the management of TRD.[52] While intravenous ketamine has evidence in both treatment-resistant major depressive disorder and bipolar depression,[48,53−57] intranasal esketamine has evidence only for the treatment of major depressive disorder. Conversely, only esketamine has FDA approval. Thus it remains unclear if intranasal esketamine can be used in the treatment of bipolar depression.

For very-short-term use, the available data demonstrates the antidepressive effects of ketamine versus esketamine treatment relative to a variety of control conditions, beginning within hours of administration and lasting up to 7 days after a single dose.[58,59] However, there is a lack of head-to-head comparisons between the two agents, which would help further clarify the comparative efficacy of these agents.

There is a real necessity in our therapeutic armamentarium for discovering and adding more effective and safer treatments for patients who have an unsatisfactory response or intolerable side effects to the current conventional antidepressive treatments.[60] To that end, all studies of ketamine and esketamine for major depression tend to enroll patients who are resistant to one or more conventional antidepressants, second-generation antipsychotics, or mood-stabilizing medications; however, the specific definitions of treatment resistance vary across studies, with the minimum number of unsuccessful trials required for trial participation to be one, two, three, or more. This indicates that the majority of trials reserve ketamine as a "last resort" treatment, so it remains unclear how ketamine may perform in individuals with nonresistant forms of unipolar or bipolar major depression.[61]

In an interesting randomized controlled trial that attempted to boost the effects of ketamine through combination with either lithium or valproate, serum lithium and valproate levels did not correlate with ketamine's antidepressant efficacy.[62,63]

Role of ketamine in electroconvulsive therapy for depression

While there is unequivocal evidence that ketamine is an effective, potent, and rapid antidepressant when used as a standalone treatment, there is conflicting evidence for ketamine augmentation of electroconvulsive therapy (ECT).[3,50,58,64−68] In a 2014 study, when equal numbers of ketamine or ECT sessions were compared to one another in the treatment of hospitalized patients, Ghasemi and colleagues[69] found that depressive symptoms significantly improved in subjects receiving the first dose of ketamine compared with the ECT group. This improvement remained significant throughout the study, and the occurrence of depressive symptoms following the second dose of ketamine was also lower than that after the second ECT.[69] This study showed that ketamine is as effective as ECT in improving depressive symptoms in patients with major depressive disorder and has more rapid antidepressant effects than ECT.[69] However, a comprehensive meta-analysis concluded that ketamine did not enhance depressive symptoms, either early in the ECT course or at the end of the study, but ketamine was found to prolong the seizure duration.[68] Thus controversy still exists regarding the definitive role of ketamine alongside ECT.

Possible mechanisms for the rapid antidepressant effects of ketamine

Part of the challenge in elucidating the comparative performance of different formulations of ketamine may lie in the lack of a clear consensus on the mechanisms underlying ketamine's therapeutic effects.[18,20] Several theoretic mechanisms have been proposed, including NMDAR modulation, GABAergic interneuron disinhibition, direct effects through ketamine metabolites (such as hydroxynorketamine [HNK]), and a plethora of downstream effects (including activation of the mechanistic target of rapamycin [mTOR], deactivation of glycogen synthase kinase 3 [GSK-3] and eukaryotic elongation factor 2 [eEF2], enhancement of brain-derived neurotrophic factor [BDNF signaling, and activation of α-amino-3-hydroxy-5-methyl-4-isoxazolepropionic acid receptors [AMPARs]).[70] These purported mechanisms may also work in a complementary and synergistic manner, which may function to enhance the antidepressive effects seen with ketamine. It is also worth noting that drugs related to ketamine—with similar receptor-binding affinities—have also been explored in order to gain an additional insight into how ketamine may work to ameliorate the symptoms of depression. These agents include scopolamine, other GluN2B-NMDAR antagonists, HNK, NMDAR glycine site modulators, NMDAR agonists (such as rapastinel), metabotropic glutamate receptor 2/3 (mGluR2/3) antagonists, $GABA_A$ receptor modulators, and drugs acting on various serotonin receptor subtypes.[70]

One of the most popular theories posits that ketamine blockade of glutamatergic neurotransmission—via antagonism of the NMDA pathway—promotes AMPAR activation.[71,72] This in turn appears to trigger a neurochemical cascade and relevant second messenger pathways that are required for several neuroplastic changes, ultimately conferring the rapid and sustained antidepressant effects of ketamine.[73,74] Although the metabolism of ketamine into HNK appears to be essential for the antidepressive properties, the receptor-binding profile of HNK appears to be independent of the NMDAR pathway and instead relies upon the AMPAR pathway.[75] HNK administration also lacks ketamine-related side effects, such as transient disorientation and dissociative experiences.[75] Taken together, this suggests the presence of a novel mechanism underlying ketamine's antidepressant properties outside the NMDAR pathway.[75] Ketamine also induces the translocation of $G_s\alpha$—a critical GTPase that activates adenylate cyclase and the cAMP-dependent signal transduction cascade—from lipid rafts to nonraft domains.[76,77] As this action requires the presence of the NMDAR, this indicates that there may be additional targets for ketamine's antidepressant properties.[76,77]

In addition to the glutamatergic pathway, previous studies have indicated ketamine activates the μ-, κ-, and δ-opioid receptors.[78–81] While the precise implications of these properties are currently under investigation, available studies indicate that the endogenous opioid system plays a role in mediating the antidepressant properties of ketamine.[49,82,83] To that end, the antidepressant effects of ketamine appear to require the activation of the opioid system, as administration of the opioid antagonist naloxone has been shown to abolish the antidepressant properties of ketamine[83]; however, another study contested these findings, claiming a lack of opioid system involvement in the antidepressant effects of ketamine.[84] Outside of depressive contexts, ketamine is also used as an adjuvant to opioid-based pharmacotherapy of pain,[85] and conversely, ketamine appears to counter opioid-induced respiratory depression,[80] which suggests that there may be a farther reaching interplay between the ketamine and opioid neurotransmitter systems outside of only depression.

Predictors of treatment response to ketamine

Among people with TRD, several studies suggest that specific features may guide the clinicians to the choice of more appropriate therapies at baseline, including ketamine.[86] The identification of depressed subpopulations that are more likely to benefit from ketamine treatment remains a priority.[87] Rong and colleagues[87] found that high body mass index and a positive family history of alcohol use disorder were the most replicated predictors of response to ketamine.

Clinical predictors may identify those more likely to benefit from ketamine within clinically heterogeneous populations. Recently, multimodal approaches have integrated genetics, proteomics, metabolomics, peripheral measures, neuroimaging, neuropsychopharmacologic challenge paradigms, and clinical perspectives to explore potential predictor, mediator, and moderator biomarkers of the rapid-acting antidepressant properties of ketamine. For example, Niciu and colleagues[88] identified several clinical correlates of greater improvement on the Hamilton Depression Rating Scale (HDRS) following ketamine administration, including higher body mass index, family history of an alcohol use disorder in a first-degree relative, and no history of suicide attempt.

Romeo and colleagues[89] found that family history of alcohol dependence, unipolar depressive disorder, and neurocognitive impairments, especially a slower processing speed, were also predictive of ketamine responsiveness. A recurring theme in the predictor literature is that there are many possible predictive factors, but few are replicated, such as personal history of alcohol dependence, no antecedent of suicide attempt, anxiety symptoms, markers of neural plasticity (slow wave activity, BDNF Val66Met polymorphism, expression of Shank 3 protein), and other neurologic factors (anterior cingulate activity, concentration of glutamine/glutamate), inflammatory factors (interleukin 6 concentration), or metabolic factors (concentration of B_{12} vitamin, D-serine, and L-serine; alterations in the mitochondrial β-oxidation of fatty acids).[89]

In another review, baseline vitamin B_{12} and folate levels were not found to be robust predictors of response to a single ketamine infusion.[90] A clinical trial by Pennybaker and colleagues[91] found that a positive family history of alcohol use disorder in a first-degree relative and greater dissociation during ketamine infusion were associated with better antidepressant response at 2 weeks, while improved measures of apparent sadness, reported sadness, inability to feel, and difficulty concentrating at day 1 correlated most strongly with antidepressant effects at 2 weeks.

In a trial of persons with bipolar depression receiving ketamine, participants with a family history of alcohol use disorder in a primary relative showed significantly greater improvement on Montgomery–Åsberg Depression Rating Scale (MADRS) scores and more attenuated psychotomimetic and dissociative scores than those with a negative family history of alcoholism.[92]

Limitations and caveats

Additional research on ketamine's long-term safety and its efficacy in reducing suicide risk is needed before clinical implementation. Larger, longer-term parallel group trials are needed to determine if efficacy can be extended and to further assess safety. Even though there are significant clinical implications of ketamine formulation in the treatment of depression, it is important to underscore that, in contrast to esketamine, there is no current FDA approval of ketamine for the treatment of major bipolar or unipolar depression.[52,93] Therefore the prescription of ketamine for the treatment of depression remains an off-label intervention. Although ketamine has demonstrated significant short-term benefits in several clinical studies, its long-term benefits remain insufficiently explored, and this may be a contributor to the current lack of FDA approval. Moreover, although ketamine represents an innovative, rapidly acting, experimental treatment for bipolar and unipolar depression, the route of administration presents a practical limitation, which has been solved to some extent with the intranasal formulation of esketamine. Furthermore, the acute response in the treatment of depression is impressive, but there remains a need for strategies that can maintain the initial response over days and weeks.

Conflict of interest statement

Dr. Zarate is listed as a coinventor on a patent for the use of ketamine in major depression and suicidal ideation; as a coinventor on a patent for the use of (2*R*,6*R*)-hydroxynorketamine, (*S*)-dehydronorketamine, and other stereoisomeric dehydro and hydroxylated metabolites of

(*R,S*)-ketamine metabolites in the treatment of depression and neuropathic pain; and as a coinventor on a patent application for the use of (2*R*,6*R*)-hydroxynorketamine and (2*S*,6*S*)-hydroxynorketamine in the treatment of depression, anxiety, anhedonia, suicidal ideation, and posttraumatic stress disorders. He has assigned his patent rights to the US government but will share a percentage of any royalties that may be received by the government. All other authors have no conflict of interest to disclose, financial or otherwise.

References

1. Herrman H, Kieling C, McGorry P, Horton R, Sargent J, Patel V. Reducing the global burden of depression: a lancet–world psychiatric association commission. *Lancet*. 2019;393:e42–e43.
2. Charlson F, van Ommeren M, Flaxman A, Cornett J, Whiteford H, Saxena S. New WHO prevalence estimates of mental disorders in conflict settings: a systematic review and meta-analysis. *Lancet*. 2019;394:240–248.
3. Corriger A, Pickering G. Ketamine and depression: a narrative review. *Drug Des Dev Ther*. 2019;13:3051–3067.
4. Albott CS, Lim KO, Forbes MK, et al. Efficacy, safety, and durability of repeated ketamine infusions for comorbid posttraumatic stress disorder and treatment-resistant depression. *J Clin Psychiatr*. 2018;79. https://doi.org/10.4088/JCP.17m11634.
5. Feder A, Parides MK, Murrough JW, et al. Efficacy of intravenous ketamine for treatment of chronic posttraumatic stress disorder: a randomized clinical trial. *JAMA Psychiatr*. 2014;71:681–688.
6. Hartberg J, Garrett-Walcott S, De Gioannis A. Impact of oral ketamine augmentation on hospital admissions in treatment-resistant depression and PTSD: a retrospective study. *Psychopharmacology*. 2018;235:393–398.
7. Krystal JH, Abdallah CG, Averill LA, et al. Synaptic loss and the pathophysiology of PTSD: implications for ketamine as a prototype novel therapeutic. *Curr Psychiatr Rep*. 2017;19:74.
8. Pradhan B, Mitrev L, Moaddell R, Wainer IW. d-Serine is a potential biomarker for clinical response in treatment of post-traumatic stress disorder using (R,S)-ketamine infusion and TIMBER psychotherapy: a pilot study. *Biochim Biophys Acta Protein Proteonom*. 2018;1866:831–839.
9. Fan W, Yang H, Sun Y, et al. Ketamine rapidly relieves acute suicidal ideation in cancer patients: a randomized controlled clinical trial. *Oncotarget*. 2017;8:2356–2360.
10. Gewandter JS, Mohile SG, Heckler CE, et al. A phase III randomized, placebo-controlled study of topical amitriptyline and ketamine for chemotherapy-induced peripheral neuropathy (CIPN): a University of Rochester CCOP study of 462 cancer survivors. *Support Care Cancer*. 2014;22:1807–1814.
11. Hardy J, Quinn S, Fazekas B, et al. Randomized, double-blind, placebo-controlled study to assess the efficacy and toxicity of subcutaneous ketamine in the management of cancer pain. *J Clin Oncol*. 2012;30:3611–3617.
12. Kannan TR, Saxena A, Bhatnagar S, Barry A. Oral ketamine as an adjuvant to oral morphine for neuropathic pain in cancer patients. *J Pain Symptom Manag*. 2002;23:60–65.

13. Lauretti GR, Lima IC, Reis MP, Prado WA, Pereira NL. Oral ketamine and transdermal nitroglycerin as analgesic adjuvants to oral morphine therapy for cancer pain management. *Anesthesiology*. 1999;90:1528–1533.
14. Lauretti GR, Gomes JM, Reis MP, Pereira NL. Low doses of epidural ketamine or neostigmine, but not midazolam, improve morphine analgesia in epidural terminal cancer pain therapy. *J Clin Anesth*. 1999;11:663–668.
15. Mercadante S, Arcuri E, Tirelli W, Casuccio A. Analgesic effect of intravenous ketamine in cancer patients on morphine therapy: a randomized, controlled, double-blind, crossover, double-dose study. *J Pain Symptom Manag*. 2000;20:246–252.
16. Zhou N, Fu Z, Li H, Wang K. Ketamine, as adjuvant analgesics for patients with refractory cancer pain, does affect IL-2/IFN-γ expression of T cells in vitro?: a prospective, randomized, double-blind study. *Medicine*. 2017;96:e6639.
17. Li L, Vlisides PE. Ketamine: 50 Years of modulating the mind. *Front Hum Neurosci*. 2016;10. https://doi.org/10.3389/fnhum.2016.00612.
18. Strasburger SE, Bhimani PM, Kaabe JH, et al. What is the mechanism of Ketamine's rapid-onset antidepressant effect? A concise overview of the surprisingly large number of possibilities. *J Clin Pharm Therapeut*. 2017;42:147–154.
19. Murrough JW, Perez AM, Pillemer S, et al. Rapid and longer-term antidepressant effects of repeated ketamine infusions in treatment-resistant major depression. *Biol Psychiatr*. 2013;74:250–256.
20. Zanos P, Gould TD. Mechanisms of ketamine action as an antidepressant. *Mol Psychiatr*. 2018;23:801–811.
21. Lüscher B, Möhler H. Brexanolone, a neurosteroid antidepressant, vindicates the GABAergic deficit hypothesis of depression and may foster resilience. *F1000Res*. 2019;8. https://doi.org/10.12688/f1000research.18758.1.
22. Kanes S, Colquhoun H, Gunduz-Bruce H, et al. Brexanolone (SAGE-547 injection) in post-partum depression: a randomised controlled trial. *Lancet*. 2017;390:480–489.
23. Kanes SJ, Colquhoun H, Doherty J, et al. Open-label, proof-of-concept study of brexanolone in the treatment of severe postpartum depression. *Hum Psychopharmacol*. 2017; 32. https://doi.org/10.1002/hup.2576.
24. Meltzer-Brody S, Colquhoun H, Riesenberg R, et al. Brexanolone injection in post-partum depression: two multicentre, double-blind, randomised, placebo-controlled, phase 3 trials. *Lancet*. 2018;392:1058–1070.
25. *Sage Slides Off a MOUNTAIN as Depression Drug Flunks Trial*; December 6, 2019. PMLive. Retrieved February 14, 2020, from http://www.pmlive.com/pharma_news/sage_slides_off_a_mountain_as_depression_drug_flunks_trial_1319276.
26. Lener MS, Kadriu B, Zarate CA. Ketamine and beyond: investigations into the potential of glutamatergic agents to treat depression. *Drugs*. 2017;77:381–401.
27. Polis AJ, Fitzgerald PJ, Hale PJ, Watson BO. Rodent ketamine depression-related research: finding patterns in a literature of variability. *Behav Brain Res*. 2019;376: 112153.
28. Pereira VS, Joca SRL, Wegener G. PS142. Ketamine is effective in an animal model of treatment resistant depression. *Int J Neuropsychopharmacol*. 2016;19:48–49.
29. Wang J, Goffer Y, Xu D, et al. A single sub-anesthetic dose of ketamine relieves depression-like behaviors induced by neuropathic pain in rats. *Anesthesiology*. 2011; 115:812–821.

30. Zhang G-F, Wang J, Han J-F, et al. Acute single dose of ketamine relieves mechanical allodynia and consequent depression-like behaviors in a rat model. *Neurosci Lett.* 2016;631:7–12.
31. Wang N, Yu H-Y, Shen X-F, et al. The rapid antidepressant effect of ketamine in rats is associated with down-regulation of pro-inflammatory cytokines in the hippocampus. *Ups J Med Sci.* 2015;120:241–248.
32. Tan S, Wang Y, Chen K, Long Z, Zou J. Ketamine alleviates depressive-like behaviors via down-regulating inflammatory cytokines induced by chronic restraint stress in mice. *Biol Pharm Bull.* 2017;40:1260–1267.
33. Clarke M, Razmjou S, Prowse N, et al. Ketamine modulates hippocampal neurogenesis and pro-inflammatory cytokines but not stressor induced neurochemical changes. *Neuropharmacology.* 2017;112:210–220.
34. Thelen C, Sens J, Mauch J, Pandit R, Pitychoutis PM. Repeated ketamine treatment induces sex-specific behavioral and neurochemical effects in mice. *Behav Brain Res.* 2016; 312:305–312.
35. Franceschelli A, Sens J, Herchick S, Thelen C, Pitychoutis PM. Sex differences in the rapid and the sustained antidepressant-like effects of ketamine in stress-naïve and "depressed" mice exposed to chronic mild stress. *Neuroscience.* 2015;290:49–60.
36. Fitzgerald PJ, Yen JY, Watson BO. Stress-sensitive antidepressant-like effects of ketamine in the mouse forced swim test. *PLoS One.* 2019;14:e0215554.
37. Irwin SA, Iglewicz A, Nelesen RA, et al. Daily oral ketamine for the treatment of depression and anxiety in patients receiving hospice care: a 28-day open-label proof-of-concept trial. *J Palliat Med.* 2013;16:958–965.
38. George D, Gálvez V, Martin D, et al. Pilot randomized controlled trial of titrated subcutaneous ketamine in older patients with treatment-resistant depression. *Am J Geriatr Psychiatr.* 2017;25:1199–1209.
39. Cullen KR, Amatya P, Roback MG, et al. Intravenous ketamine for adolescents with treatment-resistant depression: an open-label study. *J Child Adolesc Psychopharmacol.* 2018;28:437–444.
40. Wilkinson ST, Ballard ED, Bloch MH, et al. The effect of a single dose of intravenous ketamine on suicidal ideation: a systematic review and individual participant data meta-analysis. *Am J Psychiatr.* 2018;175:150–158.
41. Witt K, Potts J, Hubers A, et al. Ketamine for suicidal ideation in adults with psychiatric disorders: a systematic review and meta-analysis of treatment trials. *Aust N Z J Psychiatr.* 2020;54:29–45.
42. Witt K, Potts J, Hubers A, et al. Ketamine for suicidal ideation in adults with psychiatric disorders: a systematic review and meta-analysis of treatment trials. *Aust N Z J Psychiatr.* 2019, 0004867419883341.
43. aan het Rot M, Collins KA, Murrough JW, et al. Safety and efficacy of repeated-dose intravenous ketamine for treatment-resistant depression. *Biol Psychiatr.* 2010;67: 139–145.
44. DiazGranados N, Ibrahim LA, Brutsche NE, et al. Rapid resolution of suicidal ideation after a single infusion of an N-methyl-D-aspartate antagonist in patients with treatment-resistant major depressive disorder. *J Clin Psychiatr.* 2010;71:1605–1611.
45. Ionescu DF, Swee MB, Pavone KJ, et al. Rapid and sustained reductions in current suicidal ideation following repeated doses of intravenous ketamine: secondary analysis of an open-label study. *J Clin Psychiatr.* 2016;77:e719–725.

46. Reinstatler L, Youssef NA. Ketamine as a potential treatment for suicidal ideation: a systematic review of the literature. *Drugs R D*. 2015;15:37–43.
47. Grunebaum MF, Galfalvy HC, Choo T-H, et al. Ketamine for rapid reduction of suicidal thoughts in major depression: a midazolam-controlled randomized clinical trial. *Am J Psychiatr*. 2018;175:327–335.
48. López-Díaz Á, Fernández-González JL, Luján-Jiménez JE, Galiano-Rus S, Gutiérrez-Rojas L. Use of repeated intravenous ketamine therapy in treatment-resistant bipolar depression with suicidal behaviour: a case report from Spain. *Ther Adv Psychopharmacol*. 2017;7:137–140.
49. Williams NR, Heifets BD, Bentzley BS, et al. Attenuation of antidepressant and antisuicidal effects of ketamine by opioid receptor antagonism. *Mol Psychiatr*. 2019;24:1779–1786.
50. Xu Y, Hackett M, Carter G, et al. Effects of low-dose and very low-dose ketamine among patients with major depression: a systematic review and meta-analysis. *Int J Neuropsychopharmacol*. 2016;19. https://doi.org/10.1093/ijnp/pyv124.
51. Fava M, Freeman MP, Flynn M, et al. Double-blind, placebo-controlled, dose-ranging trial of intravenous ketamine as adjunctive therapy in treatment-resistant depression (TRD). *Mol Psychiatr*. 2018:1–12.
52. Kim J, Farchione T, Potter A, Chen Q, Temple R. Esketamine for treatment-resistant depression - first FDA-approved antidepressant in a new class. *N Engl J Med*. 2019;381:1–4.
53. Alberich S, Martínez-Cengotitabengoa M, López P, et al. Efficacy and safety of ketamine in bipolar depression: a systematic review. *Rev Psiquiatía Salud Ment*. 2017;10:104–112.
54. Bobo WV, Vande Voort JL, Croarkin PE, Leung JG, Tye SJ, Frye MA. Ketamine for treatment-resistant unipolar and bipolar major depression: critical review and implications for clinical practice. *Depress Anxiety*. 2016;33:698–710.
55. Gałuszko-Wegielnik M, Wiglusz MS, Słupski J, et al. Efficacy of Ketamine in bipolar depression: focus on anhedonia. *Psychiatr Danub*. 2019;31:554–560.
56. Ionescu DF, Luckenbaugh DA, Niciu MJ, Richards EM, Zarate CAJ. A single infusion of ketamine improves depression scores in patients with anxious bipolar depression. *Bipolar Disord*. 2015;17:438–443.
57. Kraus C, Rabl U, Vanicek T, et al. Administration of ketamine for unipolar and bipolar depression. *Int J Psychiatr Clin Pract*. 2017;21:2–12.
58. McGirr A, Berlim MT, Bond DJ, et al. A systematic review and meta-analysis of randomized controlled trials of adjunctive ketamine in electroconvulsive therapy: efficacy and tolerability. *J Psychiatr Res*. 2015;62:23–30.
59. Canuso CM, Singh JB, Fedgchin M, et al. Efficacy and safety of intranasal esketamine for the rapid reduction of symptoms of depression and suicidality in patients at imminent risk for suicide: results of a double-blind, randomized, placebo-controlled study. *Am J Psychiatr*. 2018;175:620–630.
60. Gao M, Rejaei D, Liu H. Ketamine use in current clinical practice. *Acta Pharmacol Sin*. 2016;37:865–872.
61. Aan Het Rot M, Zarate CAJ, Charney DS, Mathew SJ. Ketamine for depression: where do we go from here? *Biol Psychiatr*. 2012;72:537–547.
62. Costi S, Soleimani L, Glasgow A, et al. Lithium continuation therapy following ketamine in patients with treatment resistant unipolar depression: a randomized controlled trial. *Neuropsychopharmacology*. 2019;44:1812–1819.

63. Xu AJ, Niciu MJ, Lundin NB, et al. Lithium and valproate levels do not correlate with ketamine's antidepressant efficacy in treatment-resistant bipolar depression. *Neural Plast*. 2015;2015:858251.
64. Fond G, Loundou A, Rabu C, et al. Ketamine administration in depressive disorders: a systematic review and meta-analysis. *Psychopharmacology*. 2014;231:3663–3676.
65. Lee EE, Della Selva MP, Liu A, Himelhoch S. Ketamine as a novel treatment for major depressive disorder and bipolar depression: a systematic review and quantitative meta-analysis. *Gen Hosp Psychiatr*. 2015;37:178–184.
66. McCloud TL, Caddy C, Jochim J, et al. Ketamine and other glutamate receptor modulators for depression in bipolar disorder in adults. *Cochrane Database Syst Rev*. 2015: CD011611.
67. Caddy C, Amit BH, McCloud TL, et al. Ketamine and other glutamate receptor modulators for depression in adults. *Cochrane Database Syst Rev*. 2015. https://doi.org/10.1002/14651858.CD011612.pub2.
68. Ainsworth NJ, Sepehry AA, Vila-Rodriguez F. Effects of ketamine anesthesia on efficacy, tolerability, seizure response, and neurocognitive outcomes in electroconvulsive therapy: a comprehensive meta-analysis of double-blind randomized controlled trials. *J ECT*. 2020. https://doi.org/10.1097/YCT.0000000000000632. Publish Ahead of Print.
69. Ghasemi M, Kazemi MH, Yoosefi A, et al. Rapid antidepressant effects of repeated doses of ketamine compared with electroconvulsive therapy in hospitalized patients with major depressive disorder. *Psychiatr Res*. 2014;215:355–361.
70. Zanos P, Thompson SM, Duman RS, Zarate CA, Gould TD. Convergent mechanisms underlying rapid antidepressant action. *CNS Drugs*. 2018;32:197–227.
71. Aleksandrova LR, Phillips AG, Wang YT. Antidepressant effects of ketamine and the roles of AMPA glutamate receptors and other mechanisms beyond NMDA receptor antagonism. *J Psychiatry Neurosci*. 2017;42:222–229.
72. Zorumski CF, Izumi Y, Mennerick S. Ketamine: NMDA receptors and beyond. *J Neurosci*. 2016;36:11158–11164.
73. Evans JW, Szczepanik J, Brutsche N, Park LT, Nugent AC, Zarate CAJ. Default mode connectivity in major depressive disorder measured up to 10 Days after ketamine administration. *Biol Psychiatr*. 2018;84:582–590.
74. Maeng S, Zarate CA. The role of glutamate in mood disorders: results from the ketamine in major depression study and the presumed cellular mechanism underlying its antidepressant effects. *Curr Psychiatr Rep*. 2007;9:467–474.
75. Zanos P, Moaddel R, Morris PJ, et al. NMDAR inhibition-independent antidepressant actions of ketamine metabolites. *Nature*. 2016;533:481–486.
76. Rasenick MM, Wray N, Schappi J. Ketamine treatment translocates gsalpha from lipid rafts in cultured glial cells: similar effects to several antidepressant compounds. *FASEB J*. 2016;30:929.5.
77. Wray NH, Schappi JM, Singh H, Senese NB, Rasenick MM. NMDAR-independent, cAMP-dependent antidepressant actions of ketamine. *Mol Psychiatr*. 2019;24:1833–1843.
78. Finck AD, Ngai SH. Opiate receptor mediation of ketamine analgesia. *Anesthesiology*. 1982;56:291–297.
79. Freye E, Latasch L, Schmidhammer H, Portoghese P. Interaction of S-(+)-ketamine with opiate receptors. Effects on EEG, evoked potentials and respiration in awake dogs. *Anaesthesist*. 1994;43(Suppl 2):S52–S58.

80. Jonkman K, van Rijnsoever E, Olofsen E, et al. Esketamine counters opioid-induced respiratory depression. *Br J Anaesth.* 2018;120:1117–1127.
81. Sarton E, Teppema LJ, Olievier C, et al. The involvement of the μ-opioid receptor in ketamine-induced respiratory depression and antinociception. *Anesth Analg.* 2001;93: 1495–1500.
82. Mathew SJ, Rivas-Grajales AM. Does the opioid system block or enhance the antidepressant effects of ketamine? *Chronic Stress.* 2019;3. https://doi.org/10.1177/2470547019852073.
83. Williams NR, Heifets BD, Blasey C, et al. Attenuation of antidepressant effects of ketamine by opioid receptor antagonism. *Am J Psychiatr.* 2018;175:1205–1215.
84. Zhang K, Hashimoto K. Lack of opioid system in the antidepressant actions of ketamine. *Biol Psychiatr.* 2019;85:e25–e27.
85. Bell R, Eccleston C, Kalso E. Ketamine as an adjuvant to opioids for cancer pain. *Cochrane Database Syst Rev.* 2003:CD003351.
86. Balestri M, Calati R, Souery D, et al. Socio-demographic and clinical predictors of treatment resistant depression: a prospective European multicenter study. *J Affect Disord.* 2016;189:224–232.
87. Rong C, Park C, Rosenblat JD, et al. Predictors of response to ketamine in treatment resistant major depressive disorder and bipolar disorder. *Int J Environ Res Publ Health.* 2018;15. https://doi.org/10.3390/ijerph15040771.
88. Niciu MJ, Mathews DC, Nugent AC, et al. Developing biomarkers in mood disorders research through the use of rapid-acting antidepressants. *Depress Anxiety.* 2014;31: 297–307.
89. Romeo B, Choucha W, Fossati P, Rotge J-Y. Clinical and biological predictors of ketamine response in treatment-resistant major depression: Review. *Encephale.* 2017;43: 354–362.
90. Lundin NB, Niciu MJ, Luckenbaugh DA, et al. Baseline vitamin B12 and folate levels do not predict improvement in depression after a single infusion of ketamine. *Pharmacopsychiatry.* 2014;47:141–144.
91. Pennybaker SJ, Niciu MJ, Luckenbaugh DA, Zarate CA. Symptomatology and predictors of antidepressant efficacy in extended responders to a single ketamine infusion. *J Affect Disord.* 2017;208:560–566.
92. Luckenbaugh DA, Ibrahim L, Brutsche N, et al. Family history of alcohol dependence and antidepressant response to an N-methyl-D-aspartate antagonist in bipolar depression. *Bipolar Disord.* 2012;14:880–887.
93. The U.S. Food and Drug Administration. *FDA Approves New Nasal Spray Medication for Treatment-Resistant Depression; Available Only at a Certified Doctor's Office or Clinic.* FDA; September 11, 2019. Retrieved February 5, 2020, from http://www.fda.gov/news-events/press-announcements/fda-approves-new-nasal-spray-medication-treatment-resistant-depression-available-only-certified.

CHAPTER

How to implement a ketamine clinic

6

Ranjith Chandrasena, MD, MSc [1], Jonathan Fairbairn, BMSc, MD [1], Melody J.Y. Kang, BScH [2], Gustavo H. Vazquez, MD, PhD, FRCPC [3]

[1]*Professor, Chatham-Kent Health Alliance, Chatham, Ontario, Canada;* [2]*Master of Science, Centre for Neuroscience Studies, Queen's University, Kingston, ON, Canada;* [3]*Professor, Lead, Ketamine Clinic, Mood Disorders Outpatient Unit, Queen's University, Department of Psychiatry, Providence Care Hospital, Kingston, ON, Canada*

Background

This chapter describes the process of developing a ketamine infusion service in a public hospital setting. We outline practical issues of setting up the service, funding, staffing, physical environment, protocols for patient assessment, responding to adverse events, and monitoring of outcomes.

Administrative approval

Starting a successful ketamine infusion program at a medical institution requires a significant amount of planning and coordination. Support from an administration that has the know-how, with an ability to modify the program based on evidence-based medicine, is critical.

The chronic shortage of resources seen in institutions combined with the competing interests of different departments can often result in psychiatric services taking a low priority. For example, space in ambulatory care for the administration of ketamine and staff allocation (e.g., respiratory therapists, nurses, clerks) for mental health are often not prioritized. A successful ketamine service will first need to advocate for the existence of such a program by educating the prevalence of mental illness and the high cost associated with the current model. A multidisciplinary approach can be used to develop medical directives in keeping with the professional colleges of allied healthcare workers involved in administering ketamine. This will require relationship building across departments within the health team, such as senior decision-making staff in administration, and chiefs of other departments such as anesthesia, emergency medicine, and ambulatory care.

Developing the ketamine infusion program started with educating decision-makers on the differences between an *innovative program* and an *experimental program*. It is important for all those involved to be aware of the differences, as the

Ketamine for Treatment-Resistant Depression. https://doi.org/10.1016/B978-0-12-821033-8.00006-X

pathway to completion diverges between them. Most institutions have guidelines for implementing innovative programs through administrative structures, such as pharmacy and therapeutics committees. Experimental (research) programs, on the other hand, require a review by a research and ethics committee(s). Funding also differs between both types. Innovative programs are funded by the hospital, while funding for research at most institutions is from external sources and research funding.

During the development of our program, we used Grand-Round presentations to educate and engage hospital staff. Members of hospital administration, from ambulatory care to anesthesia, were invited to participate. These presentations create awareness of the budget-friendly treatment of treatment-refractory depression. Ultimately, ketamine infusion therapy involves significant cost-saving protocols compared to electroconvulsive therapy. There is also long-term cost reduction, with fewer emergency visits for the patient population who repetitively present with suicidal thoughts. Ideally, these presentations will introduce the safety profile of ketamine, current literature on clinical results, and patient outcomes.

Initiation of the clinic

The initial program was managed by a small number of physicians and staff. The team was trained in the administration of ketamine and the methodology for data collection, with regular meetings for knowledge and experience sharing. An initial small team assures the quality of care given to patients, as well as management of any adverse events. Once a program is established a "train-the-trainer type" educational program was applied to educate other physicians, ultimately leading to them managing their own patients who are receiving ketamine infusions, as the most responsible physician.

Ketamine is classified as a general anesthetic and thus the protocols for storage and usage of ketamine follow the norms for a drug in this class. This includes management of any adverse events that may arise during its use. Although there is no consensus on the need for an anesthetist to be present during an infusion, deploying a respiratory therapist is a budget-friendly alternative. Furthermore, a physician assistant can be the psychiatrist's eyes and ears on the floor. The quality of care can be further extended by involving dedicated nursing staff from psychiatry. Such delegation allows the psychiatrist to focus on the overall well-being of the patient without being physically present and be able to bill for the service.

Anesthesia clearance of the patients from an outpatient clinic is a compromise that allows the respiratory therapist to administer an anesthetic drug. On-call or staff psychiatrists can provide the necessary oversight during ketamine infusion. Policies and procedures need to be cross-referenced to existing protocols (e.g., cardiac arrest and other serious adverse events), albeit rare.

Patient evaluation

Patients are initially evaluated by a psychiatrist to assess their eligibility for ketamine infusions. Eligible patients are aged 18 years or older and have a diagnosis of treatment-refractory major depressive disorder or bipolar disorder according to the *Diagnostic and Statistical Manual of Mental Disorders* (Fifth Edition) (*DSM-5*) diagnostic criteria. Patients were considered treatment refractory when they did not respond adequately to two or more trials of pharmacologic treatments from different drug classes. In practice, many patients have failed three or more antidepressant trials, and/or electroconvulsive therapy/transcranial magnetic stimulation. In addition, most patients have already failed to respond to a sufficient trial of psychotherapy.

Patients were assessed using standardized clinical scales including either the Montgomery-Åsberg Depression Rating Scale (MADRS) or the Hamilton Depression Rating Scale (HAM-D). In order to be eligible for ketamine infusions, patients required an overall score of 16 points or higher on either the MADRS or HAM-D rating scale (mild-moderate severity). For clinical and research purposes, additional scales were used before and after treatment, including the Brief Psychiatric Rating Scale (BPRS), Clinician-Administered Dissociative States Scale (CADSS), Clinical Global Impression (CGI), Young Mania Rating Scale (YMRS), Sheehan Disability Scale (SDS), Snaith-Hamilton Pleasure Scale (SHAPS), and THINC-IT Cognitive Screener. The CADSS, MADRS, YMRS, and BPRS were conducted weekly in addition to the pre- and postassessments.

The exclusion criteria for ketamine infusions are based on broad safety considerations, mechanism of action, and pharmacologic/pharmacodynamic effects of ketamine (Table 6.1). Patients with a diagnosis of a mood disorder with current psychotic features are contraindicated because of the actions of ketamine on the glutaminergic system and its potential psychomimetic effects. Ketamine has the potential to lead to tolerance and addiction so patients with a comorbid alcohol or substance use disorder were contraindicated. Physical health exclusion criteria for ketamine include cardiovascular contraindications such as poorly controlled hypertension, a recent history of aortic dissection or myocardial infarction, or a diagnosis of cerebral/aortic aneurysms.[1]

Table 6.1 Exclusion criteria for ketamine infusions.

- Active substance abuse
- History of psychosis
- Active suicidal ideation/plans
- Primary diagnosis of personality disorder
- History of increased intracranial pressure
- Pregnancy or breastfeeding mother
- Uncontrolled hypertension
- Prior hypersensitivity or negative reaction to ketamine

During the assessment, a history inclusive of both prescription and over-the-counter medication use should be collected and reviewed in detail. Ketamine has few significant drug-drug interactions; however, concurrent use of monoamine oxidase inhibitors may lead to a hypertensive crisis.[2] Many ketamine studies include patients with concurrent antidepressant use including selective serotonin reuptake inhibitors and serotonin-norepinephrine reuptake inhibitors, suggesting these treatments are compatible with ketamine infusions and do not adversely affect either efficacy or tolerability.[3,4] The mechanism of subanesthetic ketamine involves multiple biological effects, one of which includes the release of glutamate, an excitatory neurotransmitter. Lamotrigine inhibits glutaminergic release, and recent studies have suggested that this may attenuate the therapeutic effects of ketamine. In addition, some low-quality studies suggest benzodiazepines may attenuate the efficacy of intravenous ketamine.[5,6]

The clinical setting for ketamine

The treatment site for ketamine infusions should include a separate waiting area and treatment area. The physical environment should be comfortable and private and should provide a quiet setting for patients and families. The treatment and recovery area should contain a comfortable chair or stretcher for the patient, as well as medical equipment including a stethoscope, sphygmomanometer, electrocardiographic monitor, pulse oximeter, and an oxygen delivery system. It is recommended that the treatment site is readily accessible to emergency equipment, supplies, and trained personnel should these resources be required.

Treatment protocol

Patients should have completed a preketamine psychiatric and general medical evaluation before their first treatment (Table 6.2). Patients were instructed to arrange alternative transportation and not to drive the day of their infusion. Before receiving ketamine, patients are asked if they have had any changes to their mental state or physical health. This should be documented and assessed by staff. Patients should

Table 6.2 Preassessment for ketamine infusions.

- Psychiatric consultation
- Anesthesiologist consultation
- Routine bloodwork (complete blood count, electrolytes, creatinine, blood urea nitrogen, and liver function tests)
- Electrocardiogram
- Height and weight
- Clinical scales
- Written informed consent

have vital signs recorded and pretreatment clinical scales completed. Height and weight should be measured, and dosing of intravenous ketamine should be verified.

When the patient is ready for the ketamine infusion, intravenous access should be established. Intravenous line access is typically maintained using a saline drip. Patients are infused through a pump, specifically CareFusion Alaris PC, with subanesthetic ketamine (0.5 mg/kg) over 40 minutes, and vital signs are monitored by a nurse regularly during the infusion, approximately every 10 minutes. Typical staffing ratios are one nurse to two or three patients. During the infusion, patients may wish to read or listen to calming music using headphones.

Following the infusion, patients should be observed, and vital signs should be monitored for 30 minutes or longer if clinically indicated. Clinical scales and reassessment of the patient's mental state should be documented after treatment as well. If patients are not able to complete the clinical scales because of treatment effects, they can be reattempted the following day. Patients should be instructed not to drive or operate heavy machinery within the next 24 hours. Patients should be instructed that if they seek medical care within the next week, they are to inform staff they received a ketamine infusion. Postketamine instructions should be supplemented by a standardized information sheet given after each treatment on discharge from the facility.

The available evidence for short-term ketamine infusions in the treatment of resistant depression on an ambulatory basis suggests it is relatively safe and well tolerated.[3,4] Adverse events of ketamine include drowsiness, dizziness, dissociative symptoms, mild tachycardia, and hypertension.[7,8] Side effects generally peak within 2 hours of receiving ketamine and resolve within 4 hours[9] A transient increase in systolic and diastolic blood pressure of up to 20 and 13 mmHg, respectively, may occur for up to 4 hours after infusion.[10] Patients should be provided with psychoeducation regarding potential side effects of ketamine before receiving treatment.

A list of adverse events, divided into serious and common, should also be kept. This is used to develop a continuously evolving risk management strategy. Using mode value rather than the mean value for the number of treatments and duration frequency has shown to be more useful for future planning and adjustment of the program.

Ketamine infusion frequency—acute and maintenance therapies

There is no clear consensus on the optimal frequency of ketamine infusions.[11] In most randomized controlled trials, infusions were provided two or three times per week over the course of 4 weeks. One study found that clinical improvement was similar among patients who received ketamine infusions twice or thrice per week.[12] There is limited evidence to indicate the number of ketamine infusions to be administered before determining that the treatment is futile.[13] However, it has been reported that some patients require up to six infusions before displaying a significant clinical response.[14]

There is limited, low-quality data regarding the benefits for and safety of patients who received maintenance ketamine infusions.[13] Given the limited published data, the risks and benefits of providing repeated long-term ketamine infusions should be carefully weighed. In addition, recreational and daily use of repeated ketamine administration report cystitis, neurocognitive deficits, and addiction concerns.[1] Following an acute course of ketamine, some centers continue maintenance ketamine treatments based on empirically derived intervals, which are patient specific. However, and according to the reported length of action for single infusion of ketamine, we are following a once-a-week infusion protocol. Standardized clinical depression rating scales such as MADRS or HAM-D are useful tools to monitor for relapse and to determine when maintenance ketamine infusions may be indicated.

Sustainability of the infusion service

A good program should involve ongoing review of the current literature, the evidence base for the program, and meaningful data collection that would allow continuous evaluation and modification. Program and protocol development allows improvement and optimization, which is different to a research protocol that is rigid and predetermined, with data analyzed at the end. By collaborating with different sites, a greater amount of data can be analyzed and will ultimately contribute scientifically to the overall improvement of all programs.

Given competing funding requirements for institution-based services, seeking out funding from outside sources (e.g., rotary and insurance funding) to be channeled back to the ketamine program through the institution increases the sustainability of such a program.

Programs that are reaching their capacity should consider partnering and referring patients to community-based services that accept private insurance. This would allow patients with private insurance to access these community-based services, allowing the program to focus on those without insurance benefits. A solid partnership between services would also allow for continuous and consistent data collection. Programs can further increase their capacity with the usage of the upcoming release of esketamine. Allowing outside care providers to administer the drug will further increase the capacity of the program. Esketamine is currently under review by Health Canada and may be available in the near future.

The primary focus of the program should be continuously increasing capacity, in order to prevent the program from hitting a roadblock due to lack of capacity, which is an eventuality as the number of patients in the maintenance program increases. An outlook period of 2–3 years should be taken into consideration.

Conclusion

This chapter summarizes the practical steps of setting up a ketamine infusion service. The success of the ketamine infusion program requires establishing links between different health care departments within the institution. In addition, sourcing funding, staffing, the physical clinic, and initiating protocols for patient assessment, response to adverse events, and monitoring of outcomes are all necessary procedures for safe and effective repeated subanesthetic ketamine administrations. Attention and care is advised when setting up similar clinics because of the open-label nature of ketamine and the current quality of evidence supporting these methods. Sharing of results and protocols would allow for similar programs to be initiated at other institutions. Partnerships with institutions that have the means and knowledge to complete sophisticated neurobiological assessments would further contribute to the advancement of knowledge. Finally, sharing of results through peer-reviewed presentations and publications will help in ketamine research and future development of other psychedelic drugs such as esketamine and psilocybin. Future recommendations include adjusting and optimizing these protocols based on subsequent research to better fit the needs of patients and undertake the safest practice possible.

References

1. Short B, Fong J, Galvez V, Shelker W, Loo CK. Side-effects associated with ketamine use in depression: a systematic review. *Lancet Psychiatry*. 2018;5(1):65–78. https://doi.org/10.1016/S2215-0366(17)30272-9. Epub 2017 Jul 27.
2. Katz RB, Toprak M, Wilkinson ST, Sanacora G, Ostroff R. Concurrent use of ketamine and monoamine oxidase inhibitors in the treatment of depression: a letter to the editor. *Gen Hosp Psychiatr*. 2018;54:62–64.
3. Andrade C. Ketamine for depression, 4: in what dose, at what rate, long, and at what frequency? *J Clin Psychiatr*. 2017;78:e852.
4. Andrade C. Ketamine for depression, 5: potential pharmacokinetic and pharmacodynamic drug interactions. *J Clin Psychiatr*. 2017;78(7):e858–e861. https://doi.org/10.4088/JCP.17f11802.
5. Albott CS, Shiroma PR, Cullen KR, et al. The antidepressant effect of repeat dose intravenous ketamine is delayed by concurrent benzodiazepine use. *J Clin Psychiatr*. 2017;78:e308.
6. Frye MA, Blier P, Tye SJ. Concomitant benzodiazepine use attenuates ketamine response: implications for large scale study design and clinical development. *J Clin Psychopharmacol*. 2015;35:334.
7. Acevedo-Diaz EE, Cavanaugh GW, Greenstein D, et al. Comprehensive assessment of side effects associated with a single dose of ketamine in treatment-resistant depression. *J Affect Disord*. 2020;263:568–575.
8. Niciu MJ, Shovestul BJ, Jaso BA, et al. Features of dissociation differentially predict antidepressant response to ketamine in treatment-resistant depression. *J Affect Disord*. 2018;232:310–315. https://doi.org/10.1016/j.jad.2018.02.049. Epub 2018 Feb 17.

9. Fond G, Loundou A, Rabu C, et al. Ketamine administration in depressive disorders; a systematic review and meta-analysis. *Psychopharmacology.* 2014;231:3663.
10. Wan LB, Levitch CF, Perez AM, et al. Ketamine safety and tolerability in clinical trials for treatment-resistant depression. *J Clin Psychiatr.* 2015;76:247.
11. Loo C. Can we confidently use ketamine as a clinical treatment for depression? *Lancet Psychiatry.* 2018;5:11.
12. Sinah JB, Fedachin M, Dalv EJ, et al. A double-blind. Randomized. Placebo-controlled. Dose-frequency study of intravenous ketamine in patients with treatment-resistant depression. *Am J Psychiatr.* 2016;173:816.
13. Sanacora G, Frye MA, McDonald W, et al. A consensus statement on the use of ketamine in the treatment of mood disorders. *JAMA Psychiatry.* 2017;74(4):399–405. https://doi.org/10.1001/jamapsychiatry.2017.0080.
14. Phillips JL, Norris S, Talbot J, et al. Single, repeated, and maintenance ketamine infusions for treatment-resistant depression: a randomized controlled trial. *Am J Psychiatr.* 2019;176(5):401–409. https://doi.org/10.1176/appi.ajp.2018.18070834.

CHAPTER

Development of new rapid-action treatments in mood disorders

7

Elisa M. Brietzke, MD, PhD [1], Rodrigo B. Mansur[3,4], Fabiano A. Gomes, MD, PhD [2,6], Roger S. McIntyre, MD, FRCPC [3,4,5]

[1]*Professor, Kingston General Hospital, Providence Care Hospital, Queen's University School of Medicine, Kingston, ON, Canada;* [2]*Kingston General Hospital, Kingston, ON, Canada;* [3]*Department of Psychiatry, University of Toronto, Toronto, ON, Canada;* [4]*Mood Disorders Psychopharmacology Unit (MDPU), University Health Network (UHN), Toronto, ON, Canada;* [5]*Brain and Cognition Discovery Foundation (BCDF), Toronto, ON, Canada;* [6]*Assistant Professor, Department of Psychiatry, Queen's University School of Medicine, Kingston, ON, Canada*

Introduction

Rapid-action antidepressants are a remarkable unmet need for patients with major depressive disorder (MDD). Since the emergence of monoaminergic antidepressants, it was rapidly recognized that, as a group, they can take up to 6 weeks to exhibit a consistent effect. Although the identification of predictors of early and sustained response to antidepressants is a "hot topic" in the medical literature, there is no reliable predictive model and trial and error using the reduction of severity of depressive symptoms is still the gold standard to identify early responders.[1,2]

The recognition of this need provoked a shift in the focus of research in depression, with several funding agencies investing in support of projects that target the development of pharmacologic and nonpharmacologic approaches that are able to reduce the severity of depression in days or weeks.[3,4]

Historically the first rapid-action treatment adopted clinically was acute sleep deprivation. A single night deprivation of sleep was first described a potentially useful treatment for depression in 1959 and was evaluated in several other studies from the 1970s.[5] A relatively recent meta-analysis on the effect of acute sleep deprivation in depression was conducted by Boland et al.[6] In this study, the overall response rate to acute sleep deprivation was 45% in randomized controlled design and around 50% in nonrandomized studies.[6] Furthermore, sleep deprivation was combined with other chronotherapeutic interventions such as light therapy and sleep phase shifting.[7] Although the mechanisms of action of the antidepressant properties of sleep deprivation were not fully elucidated, changes in gene expression of circadian genes have be postulated to play a major role.[8]

After this, electroconvulsive therapy became a treatment of choice when a rapid action is required, such as in cases of very high risk of suicide and depression with

Ketamine for Treatment-Resistant Depression. https://doi.org/10.1016/B978-0-12-821033-8.00007-1

psychotic features or catatonic features. Nevertheless, very few treatments have shown to be associated with a rapid effect. The emergence of ketamine as a rapid-action treatment was a definitive contributor for the investigation of new targets that could be used to develop novel rapid-action antidepressants.

Molecular targets for rapid-action treatments

CLOCK genes

As sleep deprivation is historically a rapid-action antidepressive treatment, genes involved in circadian rhythm have been postulated as possible mechanisms of action. CLOCK genes are those involved in the control of rhythmicity of the metabolism of cells, organs, systems, and the organism as a whole.[9] Some of these genes regulate their own expression through the production of messenger RNA (mRNA) or microRNA.[10] These genes, the mRNA derived from them, and the proteins exert functions in regulating the function of different regions of the sleep centers of the brain, chorotypes, and disruption of sleep-wake cycle in psychiatric illnesses.

Depression is strongly correlated to sleep abnormalities including insomnia and hypersomnia. Insomnia is extremely common in patients with unipolar and bipolar depression.[11] Hypersomnia is also common and happens in about 30% of the patients,[12] and combined insomnia and hypersomnia also occurs. Sleep problems are also related to poor quality of life in depression[13] and risk of relapse.[14]

Polymorphisms in CLOCK genes have been associated to MDDs, although these results are not homogeneous between studies. Their expression is altered in animal models of depression.[15] There is evidence that epigenetic changes in the CLOCK genes are related to the response to conventional antidepressants.[16]

Sleep deprivation, a historically used rapid-action treatment, modulates CLOCK genes.[17,18] The antidepressive effect of sleep deprivation is correlated with changes in transcription of CLOCK genes.[19] Although there is no evidence in humans, animal studies suggest that both sleep deprivation and ketamine administration modulate the same transcriptional biosignature of CLOCK genes.[20] One study demonstrated an overlap of 64 genes that have their transcription changed by both interventions, including Ciart, Perp2, Npas4, Dbp, and Rorb.[20]

Glutamatergic transmission

Most of the proposed mechanisms for rapid antidepressant action come from studies with ketamine and are related to its effect on glutamatergic transmission through the modulation of the *N*-methyl-D-aspartate receptor (NMDAR).[21] NMDARs are glutamatergic ion channel receptors, coactivated by glutamate and glycine, and are composed of four different subunits, with an obligatory GluN1 combined with different GluN2 subunits (GluN2A, B, C, and D). There are two other classes of glutamate ionotropic receptors, α-amino-3-hydroxy-5-methyl-4-isoxazolepropionic acid (AMPA) and kainate, that gate Na^+ and mediate fast excitatory transmission.[22] Importantly, the function of the NMDAR is tightly linked to the AMPA receptor: glutamate-AMPA-stimulated depolarization is required for opening the NMDA channel, allowing removal of Mg^{2+} that blocks the channel pore; this is required for modulators such as the antagonist ketamine to enter and block the NMDA channel.[23] Other substances may exert their action by binding to different regions such as the glycine site in the receptor.

The actions of ketamine and its stereoisomers and metabolites (i.e., (*S*)-ketamine, (*S*)-norketamine, (2*R*,6*R*)-HNK [hydroxynorketamine]), negative allosteric modulators (i.e., Ro 25–6981), and muscarinic receptor antagonists (i.e., scopolamine) are activity dependent and cause a burst of glutamate via blockade of receptors on tonic firing γ-aminobutyric acid (GABA) interneurons, resulting in disinhibition of glutamate transmission.[24] The mGluR2/3 antagonists (i.e., LY341,495 and MGS0039) also cause an increase in glutamate via blockade of presynaptic autoreceptors that provide negative feedback regulation.[25] This indirect increase in glutamate transmission is thought to be the main explanation for the downward effect, although a direct effect on glutamate transmission may also play a role.

The burst of glutamate causes activity-dependent release of brain-derived neurotrophic factor (BDNF), stimulation of TrkB-Akt, and mTORC1 signaling; these pathways lead to rapid induction of synaptic protein synthesis that is required for new synapse formation. Agents such as rapastinel, which acts as a glycinelike partial agonist, may increase synapse formation by enhancing NMDA function directly on pyramidal neurons and thereby increasing BDNF release and downstream mTORC1 signaling.[26] A requirement for mTORC1 has been demonstrated in several rapid-acting agents (i.e., blockade by the mTORC1 inhibitor rapamycin). Further support for mTORC1 is provided by evidence that a small-molecule activator of mTORC1 also produces rapid synaptic and antidepressant behavioral responses.[22]

Although chronic administration of typical monoaminergic antidepressants increases BDNF, this is limited to the expression and not activity-dependent release as observed with ketamine. Recent clinical studies demonstrate that the $GABA_A$ positive allosteric modulating agents, notably the neuroactive steroid allopregnanolone (referred to as brexanolone) and the related compound SAGE-217, also produce rapid antidepressant responses in both postpartum and general depression.[27] The intersection of these agents with the mechanisms underlying the rapid response to glutamatergic agents remains to be identified.

Cholinergic transmission

The cholinergic system has been considered as involved in the pathophysiology of depression following findings that the use of physostigmine (an acetylcholinesterase inhibitor) is able to produce depressive symptoms.[28] Physostigmine is thought to both exacerbate depressive symptoms[29] and produce depressive symptoms in bipolar patients.[30] However, one of the pitfalls of the research on the involvement of cholinergic symptoms in depression is the fact that antimuscarinic agents produce antidepressant-like effects in animals, but did not have robust antidepressant properties in humans.[31]

For example, rats from a specific lineage exhibiting supersensitivity of the muscarinic receptors display depressionlike symptoms in the presence of agents that increase cholinergic activity.[32] On the other hand, anticholinergic agents such as scopolamine attenuate depressionlike behaviors.[33] Interventional studies with scopolamine in humans produced mixed results, which are going to be described in detail in the following.

Overview of underdevelopment rapid-action treatments

Esketamine

Esketamine is the S enantiomer of ketamine and was developed by Janssen Pharmaceutica to be another option of medication for treatment-resistant depression. It was expected that esketamine be as efficient as intravenous ketamine without the complexity of the facility required to administer the second one. There is considerable disagreement in the literature about the rates of side effects of esketamine, and there is no data comparing ketamine and esketamine head-to-head. It has been studied in the treatment of treatment-resistant depression to be associated to conventional antidepressants and potentially suggesting an antisuicide effect. Esketamine should be used in an intranasal presentation.[34]

The mechanism of action of esketamine is similar to that of ketamine, with the main pharmacodynamics action being the NMDA glutamate receptor antagonist. Taken together, compared to placebo, intranasal esketamine increases the chances of response and remission in treatment-resistant depression. It appears to be safe and well tolerated in short term, but without data on long-term safety.[34]

Rapastinel

Rapastinel is an NMDAR modulator with properties of partial agonist of glycine. It has strong precognitive properties and also evidence for rapid-action antidepressive properties.[35] There are some ongoing clinical trials with rapastinel in the treatment of treatment-resistant depression.[36] Rapastinel has been used with intravenous administration in a single dose of 1, 5, 10, or 30 mg/kg. One of the limitations of this study was the large placebo effect (around 45% of reduction in Hamilton Depression Rating Scale 17 items). The results of this study suggest an antidepressive action of rapastinel, but the high placebo effect and the small sample size did not produce statistical significance.[36]

Animal and in vitro studies also support a potential antidepressant effect of rapastinel. It was seen both in behavioral tests, showing that rapastinel can attenuate depressionlike behaviors, which run in parallel with changes in neuroplasticity. For example, rapastinel is associated with increases in dendritic arborization and neurogenesis. Rapastinel is a relatively safe medication, with no serious side effects. Taken together, there are some preliminary data showing that rapastinel can modulate glutamate similar to ketamine. Unlike ketamine, it is administered in bolus and not slowly, which reduces the complexity of the facility required for its administration.

Scopolamine

Scopolamine is an anticholinergic medication that has recently been shown to also possess antidepressant effects.[29,37] Scopolamine has thought to modulate glutamatergic activity via an antagonistic effect at muscarinic receptors.[38] Scopolamine was first used to investigate the role of cholinergic system in depression, but an unexpected antidepressant effect was observed.[29] This preliminary finding supported a clinical trial with scopolamine for treatment of treatment-resistant depression. Intravenous scopolamine (dose, 4 μg/kg) was studied in a randomized clinical trial with crossover design.[29] Subjects were randomized to a placebo/scopolamine or scopolamine/placebo sequence (a series of three placebo sessions and a series of three scopolamine sessions). Sessions occurred each 3–5 days.[29] In this study, intravenous scopolamine was efficient in reducing the severity of depressive symptoms as well as anxiety after the first infusion. In addition, scopolamine was well tolerated, and no serious treatment-emergent symptom was reported. The efficacy of scopolamine in depression is in line with the hypothesis that the cholinergic system is hyperactivated in depression. The rapid improvement (immediately after the first injection up to 3 days) supports scopolamine as a rapid-acting antidepressant agent. However, limitations such as the small number of trials and crossover designs should be taken into account. An intriguing implication of these findings is the potential that the antimuscarinic activity of the tricyclic antidepressants (TCAs) can be a contributor for the relatively high antidepressive potency of those agents, especially amitriptyline. Amitriptyline is actually the TCA in which at a dose where most serotonin transporter sites are occupied, as well as most muscarinic receptors are occupied,[29] which could explain why amitriptyline is more potent than SSRIs. In clinical practice, as we see antimuscarinic effects as undesirable side effects and the amitriptyline dose is gradually titrated, it is possible that a rapid-action effect of this TCA is covert.[29]

Conclusions

Although intravenous ketamine has been shown to have a consistent rapid antidepressant effect, it is unlikely that this medication administered through an intravenous infusion itself will become a treatment for most cases of depression in clinical practice. However, one of the most valuable consequences of the introduction of intravenous ketamine is the identification of different molecular targets for the development of other rapid-action treatments, including the promising agent psilocybin. It should be highlighted that molecular targets in treatment-resistant presentations of depression are not necessarily the same for the rapid-action antidepressive effect. Future studies should elucidate the targets from ketamine that could inform the discovery of new rapid-action antidepressants, preferentially without the side effects and the complexity of the treatment.

References

1. Soares CN, Wajsbrot DB, Boucher M. Predictors of functional response and remission with desvenlafaxine 50 mg and 100 mg: a pooled analysis of randomized, placebo-controlled studies in patients with major depressive disorder. *CNS Spectrums*. May 7, 2019:1–9. https://doi.org/10.1017/S1092852919000828.
2. Masse-Sibille C, Djamila B, Julie G, Emmanuel H, Pierre V, Gilles C. Predictors of response and remission to antidepressants in geriatric depression: a systematic review. *Journal of Geriatric Psychiatry Neurology*. 2018;31(6):283–302.
3. Meltzer-Brody S, Colquhoun H, Riesenberg R, et al. Brexanolone injection in post-partum depression: two multicentre, double-blind, randomised, placebo-controlled, phase 3 trials. *Lancet*. 2018;392(10152):1058–1070.
4. Chen JCC, Sumner RL, Krishnamurthy Naga V, et al. A randomised, double-blind, active placebo-controlled, parallel groups, dose-response study of scopolamine hydrobromide (4-6 μg/kg) in patients with major depressive disorder. *Trials*. 2020;21(1):157.
5. Pflug B. The effect of sleep deprivation on depressed patients. *Acta Psychiatrica Scandinavica*. 1976;53(2):148–158.
6. Boland EM, Rao H, Dinges DF, et al. Meta-analysis of the antidepressant effects of acute sleep deprivation. *Journal of Clinical Psychiatry*. 2017;78(8):e1020–e1034.
7. Humpston C, Benedetti F, Serfaty M, et al. Chronotherapy for the rapid treatment of depression: a meta-analysis. *Journal of Affective Disorders*. 2020;261:91–102.
8. Wang XL, Yuan K, Zhang W, Li SX, Gao GF, Lu L. Regulation of circadian genes by the MAPK pathway: implications for rapid antidepressant action. *Neuroscience Bulletin*. 2020;36(1):66–76.
9. Park M, Kim SA, Yee J, Shin J, Lee KY, Joo EJ. Significant role of gene-gene interactions of clock genes in mood disorder. *Journal of Affective Disorders*. 2019;257:510–517.
10. Lamont EW, Legault-Coutu D, Cermakian N, Boivin DB. The role of circadian clock genes in mental disorders. *Dialogues in Clinical Neuroscience*. 2007;9(3):333–342.
11. Brietzke E, Vazquez GH, Kang MJY, Soares CN. Pharmacological treatment for insomnia in patients with major depressive disorder. *Expet Opinion in Pharmacotherapy*. 2019;20(11):1341–1349.
12. Grigolon RB, Trevizol AP, Cerqueira RO, et al. Hypersomnia and bipolar disorder: a systematic review and meta-analysis of proportion. *Journal of Affective Disorders*. 2019; 246:659–666.
13. Motivala SJ, Levin MJ, Oxman MN, Irwin MR. Impairments in health functioning and sleep quality in older adults with a history of depression. *Journal of the American Geriatrics Society*. 2006;54:1184–1191.
14. Kupfer DJ. Depression and associated sleep disturbances: patient benefits with agomelatine. *European Neuropsychopharmacology*. 2006;16(Suppl 5):S639–S643.

15. Christiansen SL, Bouzinova EV, Fahrenkrug J, Wiborg O. Altered expression pattern of Clock Genes in a rat model of depression. *International Journal of Neuropsychopharmacology*. December 3, 2016;19(11):pyw061.
16. Ma HY, Liu ZF, Xu YF, et al. The association study of CLOCK gene polymorphisms with antidepressant effect in Chinese with major depressive disorder. *Personalized Medicine*. 2019;16(2):115–122.
17. Hou J, Shen Q, Wan X, Zhao B, Wu Y, Xia Z. REM sleep deprivation-induced circadian clock gene abnormalities participate in hippocampal-dependent memory impairment by enhancing inflammation in rats undergoing sevoflurane inhalation. *Behavioural Brain Research*. 2019;364:167–176.
18. Goel N, Basner M, Rao H, Dinges DF. Circadian rhythms, sleep deprivation, and human performance. *Progress in Molecular Biology and Translation Science*. 2013;119: 155–190.
19. Bunney BG, Bunney WE. Mechanisms of rapid antidepressant effects of sleep deprivation therapy: clock genes and circadian rhythms. *Biological Psychiatry*. 2013;73(12): 1164–1171.
20. Orozco-Solis R, Montellier E, Aguilar-Arnal L, et al. A circadian genomic signature common to ketamine and sleep deprivation in the anterior cingulate cortex. *Biological Psychiatry*. 2017;82(5):351–360.
21. Zanos P, Gould TD. Mechanisms of ketamine action as an antidepressant. *Molecular Psychiatry*. 2018;23(4):801–811.
22. Machado-Vieira R, Henter ID, Zarate Jr CA. New targets for rapid antidepressant action. *Progress in Neurobiology*. 2017;152:21–37.
23. Aleksandrova LR, Phillips AG, Wang YT. Antidepressant effects of ketamine and the roles of AMPA glutamate receptors and other mechanisms beyond NMDA receptor antagonism. *Journal of Psychiatry Neuroscience*. 2017;42(4):222–229.
24. Kadriu B, Musazzi L, Henter ID, Graves M, Popoli M, Zarate Jr CA. Glutamatergic neurotransmission: pathway to developing novel rapid-acting antidepressant treatments. *International Journal of Neuropsychopharmacology*. 2019;22(2):119–135.
25. Duman RS, Shinohara R, Fogaça MV, Hare B. Neurobiology of rapid-acting antidepressants: convergent effects on GluA1-synaptic function. *Molecular Psychiatry*. 2019; 24(12):1816–1832.
26. Strasburger SE, Bhimani PM, Kaabe JH, et al. What is the mechanism of Ketamine's rapid-onset antidepressant effect? A concise overview of the surprisingly large number of possibilities. *Journal of Clinical Pharmacy and Therapeutics*. 2017;42(2):147–154.
27. Li YF. A hypothesis of monoamine (5-HT) - glutamate/GABA long neural circuit: aiming for fast-onset antidepressant discovery. *Pharmacology and Therapeutics*. 2020;208: 107494.
28. Furey ML, Drevets WC. Antidepressant efficacy of the antimuscarinic drug scopolamine: a randomized, placebo-controlled clinical trial. *Archives of General Psychiatry*. 2006;63:1121–1129.
29. Janowsky DS, el-Yousef MK, Davis JM. Acetylcholine and depression. *Psychosomatic Medicine*. 1974;36:248–257.
30. Risch SC, Kalin NH, Janowsky DS. Cholinergic challenges in affective illness: behavioral and neuroendocrine correlates. *Journal of Clinical Psychopharmacology*. 1981;1: 186–192.
31. Borsini F, Meli A. Is the forced swimming test a suitable model for revealing antidepressant activity? *Psychopharmacology*. 1988;94:147–160.

32. Overstreet DH, Russell RW, Hay DA, Crocker AD. Selective breeding for increased cholinergic function: biometrical genetic analysis of muscarinic responses. *Neuropsychopharmacology.* 1992;7(3):197–204.
33. Betin C, DeFeudis FV, Blavet N, Clostre F. Further characterization of the behavioral despair test in mice: positive effects of convulsants. *Physiology and Behavior.* 1982; 28(2):307–311.
34. Swainson J, Thomas RK, Archer S, et al. Esketamine for treatment resistant depression. *Expert Review of Neurotherapeutics.* 2019;19(10):899–911.
35. Moskal JR, Burgdorf JS, Stanton PK, et al. The development of rapastinel (formerly GLYX-13); a rapid acting and long lasting antidepressant. *Current Neuropharmacology.* 2017;15(1):47–56.
36. Ragguett RM, Tamura JK, McIntyre RS. Keeping up with the clinical advances: depression. *CNS Spectrums.* 2019;24(S1):25–37.
37. Drevets WC, Furey ML. Replication of scopolamine's antidepressant efficacy in major depressive disorder: a randomized, placebo-controlled clinical trial. *Biological Psychiatry.* 2010;67:432–438.
38. Duman RS. Pathophysiology of depression and innovative treatments: remodeling glutamatergic synaptic connections. *Dialogues in Clinical Neuroscience.* 2014;16:11–27.

Further reading

1. Gillin JC, Sutton L, Ruiz C, et al. The effects of scopolamine on sleep and mood in depressed patients with a history of alcoholism and a normal comparison group. *Biological Psychiatry.* 1991;30(2):157–169.

CHAPTER

Closing remarks

8

Elisa M. Brietzke, MD, PhD [1], Carlos A. Zarate, MD [2], Gustavo H. Vazquez, MD, PhD, FRCPC [3]

[1]*Professor, Kingston General Hospital, Providence Care Hospital, Queen's University School of Medicine, Kingston, ON, Canada;* [2]*Chief, Experimental Therapeutics and Pathophysiology Branch, Section Neurobiology and Treatment of Mood Disorders, Division of Intramural Research Program, National Institute of Mental Health, Bethesda, MD, United States;* [3]*Professor, Lead, Ketamine Clinic, Mood Disorders Outpatient Unit, Queen's University, Department of Psychiatry, Providence Care Hospital, Kingston, ON, Canada*

Introduction

The robust, rapid, and sustainable antidepressant effect of ketamine provoked a revolution in the treatment of depressive episodes, especially for patients with treatment-resistant presentations of unipolar depression (treatment-resistant depression [TRD]). It is an exciting moment not only for the clinicians who deeply need novel and better options for the treatment of depression but also for the researchers. The efficacy of a drug with complex actions such as intravenous ketamine offered insights about the pathophysiology of depression and provided a new stimulus for investigation of molecular targets that go beyond the monoaminergic system.[1]

An interesting aspect of the introduction of ketamine in the clinical practice is that this medication is arriving to the clinicians at a moment when the methods of investigation are advanced enough to support a detailed evaluation of its effects. For example, data on efficacy, safety, and tolerability were almost immediately paired with those obtained by neuroimaging, molecular biology, cognitive assessment, and multi-omics platforms.

In spite of that, there is still a lack of robust data in the literature on different clinical and neurobiological actions of intravenous ketamine. In this section, we examine the consolidated knowledge on intravenous ketamine for the treatment of depression and provide to the reader a summary of the current challenges in the field, which should be addressed and expanded in future research.

State of the art and current challenges

With the accumulation of studies in the efficacy and safety of tolerability for depression, in the past few years, the clinicians had available some papers of

Ketamine for Treatment-Resistant Depression. https://doi.org/10.1016/B978-0-12-821033-8.00008-3

consensus in the use of ketamine. For example, the American Psychiatric Association (APA) launched a consensus statement about ketamine in mood disorders in 2017.[2] More recently, the Canadian Network for Mood and Anxiety Treatments (CANMAT) also organized a task force to make recommendations on the use of ketamine in Canada (Lam R, in preparation). Both consensuses highlight the robust data on ketamine and the remarkable advance that intravenous ketamine represented in the field of mood disorders. However, they also show the caveats and challenges in the implementation of the research findings in clinical practice. The caution of the experts invited to elaborate the consensuses contrasts with the coverage of ketamine by the media, which not always provided a balanced view of the benefits and limitations of this drug in the treatment of depression.[3] The optimism of ketamine for treatment-resistant presentations of depression brought to an increase of interest of clients, advocacy organizations, hospitals, and doctors on this drug. The use of ketamine then became spread, including the off-label use without enough data to support its use in other psychiatric illnesses or in different contexts.[4]

Some of the challenges in the translation of the findings from the clinical trials on intravenous ketamine can be summarized in the following sections.

Limitations of knowledge

Lack of uniformity of the definition of TRD: Most trials with ketamine use the definition of TRD as the lack of response to at least two antidepressants in appropriate dose and for sufficient time to observe a response. However, in clinical practice, the adoption of a broader definition of TRD will increase the number of patients who would be candidates to ketamine treatment, challenging the existing resources to offer this treatment. The use of ketamine in ultraresistant cases of depression tends to not show comparable efficacy of less resistant patients. In the translation of the level of resistance to clinical practice, what is usually done is a balance of the risks, costs, and benefits of the treatment with intravenous ketamine for each patient, taking into account the number of medications already tried, the tolerability of different agents, the psychiatric and medical comorbidity, previous response to nonpharmacologic treatments including neuromodulatory approaches, and potential reasons for resistance.

Lack of comparison of intravenous ketamine with other approved interventions for TRD: There is no data in the literature on the comparison of intravenous ketamine and electroconvulsive therapy or transcranial magnetic stimulation or even with potentiation strategies such as the use of lithium.

Lack of enough knowledge in the efficacy of intravenous ketamine in other psychiatric illnesses: Although there are some preliminary data on the use of ketamine for the treatment of depressive episodes associated with bipolar disorder,[5] posttraumatic stress disorders,[6] and even schizophrenia,[7] these studies are small and we believe there is not enough evidence to indicate intravenous ketamine as a treatment for other conditions but rather for depressive episodes associated with major depressive disorder.

Challenges in knowledge translation

Lack of uniformity in the characteristics of facilities administering intravenous ketamine: Although this aspect was not systematically evaluated in the literature, it is possible to identify different levels of care when facilities that offer intravenous ketamine are compared to one another. For example, in some of the facilities, there is always an anesthesiologist present, some have a psychiatrist with different levels of training in life support, and some others have only other health professionals, such as trained nursing staff.

Lack of uniformity in the routes of administration: One of the most important challenges in the treatment of ketamine is its use by psychiatrists who are not necessarily linked to academic centers or who do not use the drug according to the research protocols. This is the case, for example, with the other routes of administration, such as oral ketamine.[8] Although consensus on ketamine does not recommend its use,[2] there is reasonable evidence supporting oral administration as potentially useful clinically, both in cases in which the intravenous route is not possible and potentially as a maintenance treatment. However, a systematic review has shown that oral ketamine is unlikely to have a robust effect as a rapid action treatment as well as an antisuicidal effect.[8]

No standardization of training of professionals to administer ketamine for TRD: There is no guideline about who should be trained to work in ketamine clinics and how the training should be performed. The APA consensus establishes that the doctors who deliver the treatment should be prepared to deal both with clinical and psychiatric emergencies. It is expected that they are familiar with procedures to manage potential cardiovascular events, being recommended that the professional has Advanced Cardiac Life Support certification. In the same way, the doctor should be familiar with the management of psychomotor agitation and emergence of suicidal ideation. In addition, the facility should be able to provide rapid follow-up evaluations for patients who exhibit treatment-emergent psychiatric symptoms.[2] In addition, clinicians should develop some expertise and experience on the methods of administrations of ketamine, which can be subject to national or local regulation (Table 8.1).

Conclusion

It is expected that the next few years will bring data on ketamine use as a maintenance therapy for depression or that at least other medications that act on the same targets, especially glutamatergic agents, become available as a long-term treatment for patients who respond acutely.[9] It is also expected that the targets that emerged as relevant from the studies with ketamine will be converted in novel avenues of treatment, renewing the hopes of patients and families affected by depression.

Table 8.1 Consolidated knowledge in intravenous ketamine for the treatment of patients with treatment-resistant depression and aspects to be established.

Well established	Acute efficacy for TRD	Based on open-label RCTs and meta-analysis
	Overall safety	Severe side effects are rare and most patients present with mild and transient dissociative experiences or increases in blood pressure
	Nonmonoaminergic mechanism of action	Glutamate/GABAergic modulation
Reasonably established	Acute antisuicidal effect	Evidence for some independent antisuicidal effect
	Predictors of response	High BMI, history of alcohol abuse, and cognitive decline
	Neurotrophic action	Preliminary evidence of neurotrophic and antiinflammatory effect
To be established	Maintenance treatment for TRD	There are no data on the safety of the continued use of intravenous ketamine. Data from addiction to ketamine raise concerns on long-term safety.
	Efficacy in other conditions	There are preliminary data on BD and PTSD and case reports on depression in schizophrenia.
	Efficacy and safety of other routes of administration	There is some evidence supporting oral use, although data from efficacy are not as robust as the intravenous pathway. In addition, in theory intravenous route is safer, as the infusion can be interrupted if side effects emerge.

BD, *bipolar disorder;* BMI, *body mass index;* GABAergic, *γ-aminobutyric acid mediated;* PTSD, *posttraumatic stress disorder;* TRD, *treatment-resistant depression.*

References

1. Gerhard DM, Wohleb ES, Duman RS. Emerging treatment mechanisms for depression: focus on glutamate and synaptic plasticity. *Drug Discovery Today.* 2016;21(3):454–464.
2. Sanacora G, Frye MA, McDonald W, et al. American psychiatric association (APA) council of research task force on novel biomarkers and treatments. *JAMA Psychiatry.* 2017; 74(4):399–405.
3. Zhang MW, Hong YX, Husain SF, Harris KM, Ho RC. Analysis of prints news media framing of ketamine treatment in the Unites States and Canada from 2000 to 2015. *PLoS.* 2017;12(2):e0173202.
4. Wilkinson ST, Toprak M, Turner MS, Levine SP, Katz RB, Sanacora. A survey of the clinical, off-label use of ketamine as a treatment for psychiatric disorders. *The American Journal of Psychiatry.* 2017;174(7):695–696.

5. Galuszko-Wegielnik M, Wiglusz MS, Slupski J, et al. Efficacy of ketamine in bipolar depression: focus on anhedonia. *Psychiatria Danubina*. 2019;31(Suppl 3):554–560.
6. Liriano F, Hatten C, Schwartz TL. Ketamine as treatment for post-traumatic stress disorder: a review. *Drugs Context*. 2019;8:212305.
7. Bartova L, Papageorgiou K, Milenkivic I, et al. Rapid antidepressant effect of S-ketamine in schizophrenia. *European Neuropsychopharmacology*. 2018;28(8):980–982.
8. Rosenblat JD, Carvalho AF, Li M, Lee Y, Subramanieapillai M, McIntyre RS. Oral ketamine for depression: a systematic review. *Journal of Clinical Psychiatry*. 2019;80(3): 18r12475.
9. Lener MS, Kadriu B, Zarate Jr CA. Ketamine and beyond: investigations into the potential of glutamatergic agents to treat depression. *Drugs*. 2017;77(4):381–401.

Index

'*Note*: Page numbers followed by "f" indicate figures and "t" indicate tables.'

A

α-Amino-3-hydroxy-5-methyl-4-isoxazolepropionic acid receptors (AMPARs), 13f, 16
Acetyl-L-carnitine (ALC), 52
Adjunctive antidepressants, 45–46
Altered states of consciousness (ASC), 5–6
Amantadine, 49
γ-Aminobutyric acid (GABA), 15
γ-Aminobutyric acid (GABA)-mediated (GABAergic) deficit hypothesis, 117–118
Anesthesia clearance, 132
Anterior cingulate cortex (ACC), 15
Anticonvulsants, suicidal ideation/behavior, 96
Antidepressants
  ketamine. *See* Ketamine
  suicidal ideation/behavior, 92–94, 102
  switching strategies, 42–45
Antidepressant Treatment History Form (ATHF), 35t–36t, 37
Antiinflammatory agents, 50
Anxiolytics, 48, 98
Atypical antipsychotics, 46–47
  aripiprazole, 46–47
  brexpiprazole, 46–47
  monotherapy, 47
  placebo controlled trials, 46
  vs. typical antipsychotics, 46
Augmentation strategies, 43t–44t
Ayahuasca, 58–59

B

Benzodiazepines, 48
Bipolar disorder (BD), 86, 87t
Brain-derived neurotrophic factor (BDNF), 13f, 17, 40
Brexanolone, 117–118, 141
Brief Psychiatric Rating Scale (BPRS), 133
Bupropion, 42–45
Butterfield's concept, 2
Butyrophenone, 8

C

Canadian Network for Mood and Anxiety Treatments (CANMAT), 147–148
Celecoxib, 50
Children's Depression Rating Scale Revised (CDRS-R), 119
Cholinergic transmission, 141–142
Clinician-Administered Dissociative State Scale (CADSS), 7, 133
Clock genes, 140
Clozapine, 91, 96–97
Cognitive behavior analysis system of psychotherapy (CBASP), 57
Cognitive behavior therapy (CBT), 100, 101t
Continuation electroconvulsive therapy (C-ECT), 53
Continuously theta burst stimulation (cTBS), 54–55
Cothymia, 1
Cyclohexamine (CI400), 3
Cyclooxygenase 2 (COX-2) inhibitor, 50

D

Deep brain stimulation (DBS), 55–56
Depression, 3
Depression rating scale, 34
Diagnostic and Statistical Manual of Mental Disorders (DSM) classification system, 1
Dialectical behavior therapy (DBT), 100
Dimethyltryptamine, 8
Dissociation
  altered states of consciousness (ASC), 5–6
  Berrios report, 5
  dissociative reaction, 5–6
  electrophysiologic mechanism, 4–5
  psychodynamic model of mental illness, 6
Dissociative anesthesia, 3–5
  cyclohexamine, 3
  dissociation, 4
  dissociative, 4
  emergence reaction, 4
  phencyclidine (PCP), 3
Dissociative disorders, 9
Dizocilpine, 8
Dopaminergic agents, 49
Droperidol, 8

E

Electroconvulsive therapy (ECT), 53
  ketamine, 120–121
  suicidal ideation/behavior, 100
Esketamine, 59–60, 142
Estrogen replacement therapy (ERT), 51

Eukaryotic elongation factor 2 (eEF2), 13f, 18
European Staging Model (ESM), 35t–36t, 37

F

Frank emergence delirium, 4

G

Gabapentin, 50
Glutamate, 15
Glutamate dysregulation, 40
Glutamatergic transmission, 140–141
Glutamine, 15
Glycogen synthase kinase 3 (GSK-3), 13f, 19

H

Hamilton Depression Rating Scale (HAM-D), 34, 133
Hippocratic triangle, 2
Hydroxynorketamine (HNK), 121–122
Hypersomnia, 140

I

Indole hallucinogen hypothesis, schizophrenia, 8
Infliximab, 50
Insomnia, 140
Insulin like growth factor 2 (IGF-2), 13f, 19
Intermittently theta burst stimulation (iTBS), 54–55
Interpersonal psychotherapy (IPT), 57
InterSePT trial, 91
Intranasal esketamine, 120
Intravenous ketamine, 150t
  administering facilities, 149
  administration routes, 149
  challenges, 147–148
  efficacy, 148
  lack of comparison, 148
  professional training, 149

K

Ketamine
  antidepressant properties, 118–119
    animal models, 118–119
    open-label and nonrandomized studies, 119
  in electroconvulsive therapy, 120–121
  historical background, 117–118
  history
    depression, 3–5
    functional approach, 2
    Hippocratic triangle, 2
    mental illness, 2–3
    presentist and hagiographic approach, 2
  intravenous, 150t
    administering facilities, 149
    administration routes, 149
    challenges, 147–148
    efficacy, 148
    lack of comparison, 148
    professional training, 149
  limitations and caveats, 123
  murine models, 118–119
  NMDA mechanism, 117–118
  psychotomimetic effects, 7
  randomized controlled trials, 117–118, 120
  rapid antidepressant effects
    glutamatergic pathway, 121–122
    hydroxynorketamine (HNK), 121–122
    molecular neuroplasticity. *See* Molecular neuroplasticity
    opioid system, 122
    vs. sustained effects, 21–23
    theoretic mechanisms, 121
  rat models, 118–119
  response rates, 13–14
  revolutionary halo, 1–2
  suicidal ideation/behavior, 98–100, 99t
  suicidal thought reduction, 119
  treatment response predictors, 122–123
Ketamine infusion service
  acute and maintenance therapies, 135–136
  administrative approval, 131–132
  adverse events, 135
  clinical setting, 134
  clinic initiation, 132
  exclusion criteria, 133, 133t
  experimental program, 131–132
  innovative program, 131–132
  intravenous line access, 135
  multidisciplinary approach, 131
  patient evaluation, 133–134
  patient's mental state reassessment, 135
  preassessment, 134t
  presentations, 132
  program and protocol development, 136
  sustainability, 136
  train-the-trainer type educational program, 132
  treatment protocol, 134–135
  treatment site, 134

L

Lamotrigine, 49, 134
Lithium, 19, 47, 94–95
Looping effect, 7
Lysergic acid diethylamide (LSD), 58

M

Magnetic seizure therapy (MST), 53–54
Major depression, suicidal risks, 86, 87t
Major depressive disorder (MDD)
  ketamine. *See* Ketamine
  neuroplasticity, 14
  treatment resistance, 33
Massachusetts General Hospital Staging (MGH-S) model, 35t–36t, 37
Mature brain-derived neurotrophic factor (mBDNF), 17
Maudsley Staging Model (MSM), 35t–36t
Metabotropic glutamate receptor 4 (GRM4), 21
3,4-Methylenedioxymethamphetamine (MDMA), 58
MicroRNA expression, 21
Mindfulness-based cognitive therapy (MBCT), 57
Minocyclin, 50
Mirtazapine, 42–46
MK-801, 8
Modafinil, 48–49
Molecular neuroplasticity, 14
  α-amino-3-hydroxy-5-methyl-4-isoxazolepropionic acid receptors (AMPARs), 16
  gama-aminobutyric acid (GABA), 15
  gamma oscillations, 23
  glutamate, 15
  glutamate burst, 23
  glutamine, 15
  limitations, 22–23
  MAPK/ErK inhibition, 20–21
  microRNA expression, 21
  opioid system activation, 23
  target of rapamycin (TOR) signaling pathway activation, 16–20
    brain-derived neurotrophic factor (BDNF), 17
    depression models, 20
    eukaryotic elongation factor 2 (eEF2), 18
    glycogen synthase kinase 3 (GSK-3), 19
    limitations, 19–20
    p70 ribosomal S6 kinase (p70S6K), 19
    protein kinase, 16
    proteins involved, 16
    VGF, 18
Monoamine hypothesis, 40
Monoamine oxidase inhibitors (MAOIs), 37, 45
Monoaminergic antidepressants, 139
Montgomery-°Asberg Depression Rating Scale (MADRS), 34, 123, 133
Mood-stabilizing anticonvulsants, 96
Muscarinic cholinergic receptor antagonists, 51

N

N-acetylcysteine (NAC), 52
Neuroactive steroid allopregnanolone, 141
Neuroleptoanalgesia, 8
Neuromodulation, treatment-resistant depression (TRD), 44t
  deep brain stimulation (DBS), 55–56
  electroconvulsive therapy (ECT), 53
  magnetic seizure therapy (MST), 53–54
  repetitive transcranial magnetic stimulation (rTMS), 54
  theta burst stimulation (TBS), 54–55
  transcranial direct current stimulation (tDCS), 56–57
  trigeminal nerve stimulation (TNS), 56
  vagus nerve stimulation (VNS), 56
Neuroplasticity hypothesis, 14
Nicotinic antagonists, 51
N-methyl-D-aspartate (NMDA) receptor, 13–14, 40, 59–60, 117, 140–141
Nutraceuticals, 52

O

Olanzapine, 91
Opioids, 51–52

P

Personality disorders, 87
Phencyclidine (PCP), 3, 6
Physostigmin, 141–142
Pindolol, 46
Polymorphisms, CLOCK gene, 140
Postsynaptic density protein 95 (PSD-95), 13f, 19
Pramipexole, 49
Precursor protein brain-derived neurotrophic factor (proBDNF), 17
Pregabalin, 50
p70 ribosomal S6 kinase (p70S6K), 19
Psychedelics
  antidepressant effects, 57–58
  ayahuasca, 58–59
  lysergic acid diethylamide (LSD), 58
  3,4-methylenedioxymethamphetamine (MDMA), 58
  psilocybin, 58
Psychodynamic model of mental illness, 6
Psychostimulants, 48–49
Psychotherapy
  suicidal ideation/behavior, 100
  treatment-resistant depression (TRD), 57

Q

Quetiapine, 47

R

Randomized controlled trials (RCTs)
  antidepressants, suicidal ideation/behavior, 93–94
  ketamine, 120
Rapastinel, 142–143
Rapid-action antidepressants
  cholinergic transmission, 141–142
  clock genes, 140
  esketamine, 142
  glutamatergic transmission, 140–141
  molecular targets, 140–142
  rapastinel, 142–143
  scopolamine, 143
Reboxetine, 46
Repetitive transcranial magnetic stimulation (rTMS), 54
Riluzole, 49
Risperidone, 47

S

S-adenosylmethionine (SAMe), 52
SAGE-217, 141
Scopolamine, 142–143
Second-generation antipsychotics (SGAs), 97
Sedatives, suicidal ideation/behavior, 98
Selective serotonin reuptake inhibitors (SSRIs), 37, 42–45
Serotonin-norepinephrine reuptake inhibitors (SNRIs), 37, 42–45
Sex steroids, 51
Sirukuma, 50
Sleep deprivation, CLOCK genes, 140
Stimulant like drugs, 48–49
Subanesthetic ketamine, 134
Subcallosal cingulate gyrus deep brain stimulation (SCC DBS), 55
Suicidal ideation/behavior
  anticonvulsants, 96
  antidepressants, 92–94, 102
  antipsychotics
    clozapine, 96–97
    first-generation neuroleptic drugs, 96
    second-generation antipsychotics (SGAs), 97
  anxiolytics and sedatives, 98
  ketamine, 98–100, 99t
  lithium, 94–95, 104
  prevalence, 85
  preventive challenges, 90–92, 90t
    adequate sampling, 91
    ethical challenges, 90–91
    exposure time, 91
    psychiatric treatments, 92
    severity/riskiness, 91–92
  psychotherapy, 100, 104
  risk factors
    bipolar disorder, 88
    depressed mood, 87
    depression, 88
    major mood disorders, 87–88, 89t
    mood disorder, 87–88
    morbid state, 87
    personality disorders, 87
  US adult population, 85, 86t
Suicidal risks
  general population, 85
  international suicide rate, 85
  psychiatric disorders, 86
  US adults, 85, 86t
Synthetic cannabinoids, 8–9

T

Target of rapamycin (TOR) signaling pathway activation, 16–20
  brain-derived neurotrophic factor (BDNF), 17
  depression models, 20
  eukaryotic elongation factor 2 (eEF2), 18
  glycogen synthase kinase 3 (GSK-3), 19
  limitations, 19–20
  p70 ribosomal S6 kinase (p70S6K), 19
  protein kinase, 16
  proteins involved, 16
  vascular endothelial growth factor (VEGF), 18
Temperament and Character Inventory, 39–40
Testosterone therapy, 51
Thase-Rush Staging Model (TRSM), 35t–36t, 37
Theta burst stimulation (TBS), 54–55
Thyroid hormone (T_3), 47
Tocilizumab, 50
Topiramate, 50
Train-the-trainer type educational program, 132
Transcranial direct current stimulation (tDCS), 56–57
Treatment resistant depression (TRD)
  adequate response, 34
  adequate trial, 34
  antidepressant switching strategies, 42–45, 42t
  augmentation strategies, 43t–44t
  clinical risk factors, 39–40
  combination and adjunctive therapies
    anticonvulsants, 49–50

antiinflammatory agents, 50
anxiolytics, 48
atypical antipsychotics, 46–47
bupropion and mirtazapine, 45–46
dopaminergic agents, 49
lithium, 47
mianserin and reboxetine, 46
muscarinic cholinergic receptor antagonists, 51
nicotinic antagonists, 51
nutraceuticals, 52
opioids, 51–52
pindolol, 46
psychostimulants and stimulant like drugs, 48–49
sex steroids, 51
thyroid hormone (T_3), 47
definition, 33–34
diagnosis, 38–39
differential diagnosis, 38–39
esketamine, 59–60
etiopathology
brain-derived neurotrophic factor (BDNF), 40
childhood adversity, 39–40
comorbid psychiatric disorders, 39
glutamate dysregulation, 40
hippocampal volume, 40
neurotransmitter dysregulation, 40
neurotransmitter gene expression, 40
stressful life events, 39–40
intravenous ketamine, 150t
administering facilities, 149
administration routes, 149
challenges, 147–148
efficacy, 148
lack of comparison, 148
professional training, 149
ketamine, 59
medical comorbidities, 38–39
medication nonadherence, 38
monotherapy, 41
neuromodulation
deep brain stimulation (DBS), 55–56
electroconvulsive therapy (ECT), 53
magnetic seizure therapy (MST), 53–54
repetitive transcranial magnetic stimulation (rTMS), 54
theta burst stimulation (TBS), 54–55
transcranial direct current stimulation (tDCS), 56–57
trigeminal nerve stimulation (TNS), 56
vagus nerve stimulation (VNS), 56
neuromodulation treatments, 44t
pharmacokinetics and pharmacodynamics, 38
polytherapy, 41
prevalence, 33, 37–38
psychedelics
antidepressant effects, 57–58
ayahuasca, 58–59
lysergic acid diethylamide (LSD), 58
3,4-methylenedioxymethamphetamine (MDMA), 58
psilocybin, 58
psychiatric comorbidities, 39
psychotherapy, 57
randomized controlled trial (RCT), 38
staging models, 34–37, 35t–36t
supratherapeutic doses, 38
unipolar depression, 147
Tricyclic antidepressants (TCAs), 37, 45, 143
Trigeminal nerve stimulation (TNS), 56
Tropomyosin receptor kinase B (TrkB), 13f, 17

V

Vagus nerve stimulation (VNS), 56
Vascular endothelial growth factor (VEGF), 20–21
Vortioxetin, 45

Y

Young Mania Rating Scale (YMRS), 133

Z

Ziprasidone, 47

Printed in the United States
By Bookmasters